Table of Contents

Introduction

Part I: Understanding Disasters
1. Types of Natural Disasters
2. Man-made Disasters and Emergencies
3. Assessing Your Risks and Vulnerabilities

Part II: Essential Survival Skills
4. Shelter: Finding and Building Safe Spaces
5. Water: Sourcing, Purification, and Conservation
6. Fire: Starting, Maintaining, and Using Fire Safely
7. Food: Foraging, Hunting, and Food Preservation Techniques
8. First Aid and Medical Care in Emergencies
9. Navigation and Signaling Techniques
10. Self-Defense and Security Measures

Part III: Emergency Preparedness Kit Assembly
11. The Importance of an Emergency Kit
12. Water and Food Supplies for Your Kit
13. First Aid and Medical Supplies
14. Shelter and Bedding Essentials
15. Communication and Lighting Devices
16. Hygiene and Sanitation Items
17. Tools and Multipurpose Gear
18. Customizing Your Kit for Specific Needs

Part IV: Disaster Planning and Preparation
19. Developing a Family Emergency Plan
20. Establishing Communication Protocols

21. Fortifying Your Home for Disasters
22. Vehicle Preparedness and Maintenance
23. Financial Readiness and Document Protection
24. Psychological Preparation and Mental Resilience
25. Community Involvement and Resource Sharing

Part V: Responding to SHTF Scenarios
26. Immediate Actions During a Disaster
27. Evacuation Procedures and Bug-Out Strategies
28. Sheltering in Place Effectively
29. Long-Term Survival in Catastrophic Events
30. Post-Disaster Recovery and Rebuilding

Conclusion: Maintaining a Prepared Mindset

Introduction

Imagine waking up to find your world turned upside down by a devastating earthquake, a raging wildfire, or a catastrophic hurricane. In an instant, your daily life is shattered, and you're left struggling to survive without access to basic necessities like food, water, and shelter. This is the harsh reality for millions of people each year who find themselves caught in the path of natural disasters and man-made emergencies.

The question is, are you prepared to face such a scenario? Do you have the knowledge, skills, and resources to keep yourself and your loved ones safe during a crisis? If your answer is no, or if you're unsure, then this book is for you.

In "How to Prepare for Disasters," you'll discover a comprehensive guide to disaster preparedness that will empower you to face any emergency with confidence and resilience. You'll learn how to assess your risks, acquire essential survival skills, assemble a well-stocked emergency kit, and create a foolproof disaster plan.

But this book is more than just a collection of tips and checklists. It's a roadmap to peace of mind, knowing that you've taken the necessary steps to protect yourself and your family from the unexpected. By implementing the strategies and techniques outlined in these pages, you'll gain the ability to adapt to any situation, overcome challenges, and emerge stronger on the other side.

Imagine the sense of security and empowerment you'll feel when you have the tools and knowledge to handle any crisis that comes your way. Picture the relief on your loved ones' faces when they see you taking charge and leading them to safety. And think of the pride you'll feel in knowing that you've done everything in your power to ensure their well-being.

Don't wait until disaster strikes to start preparing. Take action now and invest in your future safety and resilience. With "How to Prepare for Disasters" as your guide, you'll be ready to face whatever challenges life throws your way. So, let's dive in and start building your disaster preparedness toolkit today!

Part I
Understanding Disasters
Types of Natural Disasters

Natural disasters can strike anywhere, at any time, leaving devastation in their wake. To effectively prepare for these events, you must first understand the types of disasters that could affect your area and the potential impacts they can have on your life.

Let's dive into some of the most common natural disasters:

a. Earthquakes
- Caused by the sudden release of energy in the Earth's crust, resulting in seismic waves
- Can trigger landslides, tsunamis, and fires
- Damage to buildings, infrastructure, and utilities can be severe
- Injuries and fatalities can occur due to falling debris, collapsing structures, and secondary hazards

Imagine you're going about your daily routine when suddenly, the ground beneath your feet starts to shake violently. The walls crack, objects fall from shelves, and you struggle to maintain your balance. This is the reality of an earthquake, a powerful force of nature that can change your life in a matter of seconds.

b. Hurricanes and Tropical Storms
- Characterized by strong winds, heavy rainfall, and storm surges

- Can cause flooding, wind damage, and power outages
- Coastal areas are particularly vulnerable, but inland regions can also be affected
- Evacuation may be necessary in high-risk areas

Picture a mighty storm brewing over the ocean, gathering strength as it approaches land. As the hurricane makes landfall, fierce winds batter your home, ripping shingles from the roof and toppling trees. Torrential rain floods the streets, making roads impassable and leaving you stranded.

c. Tornadoes
- Rapidly rotating columns of air extending from a thunderstorm to the ground
- Can cause severe wind damage, destroy buildings, and hurl debris
- Often accompanied by large hail and heavy rain
- Can occur with little warning, making immediate action crucial

You're watching the news when a tornado warning is issued for your area. You scramble to gather your family and seek shelter in a safe room or basement. The roar of the wind is deafening as the twister passes nearby, leaving a path of destruction in its wake.

d. Wildfires
- Uncontrolled fires that burn through forests, grasslands, and communities
- Can spread quickly, driven by wind and dry conditions
- Pose risks to life, property, and air quality

- May require evacuation and can disrupt daily life for weeks or months

Imagine a small spark in a tinder-dry forest, quickly growing into a raging inferno. The smoke fills the air, making it difficult to breathe as the flames consume everything in their path. You're forced to flee your home, uncertain of what you'll find when you return.

e. Floods
- Can result from heavy rainfall, rapid snowmelt, or storm surges
- Cause damage to homes, businesses, and infrastructure
- Pose risks of drowning, electrocution, and water contamination
- May require evacuation and can disrupt transportation and utilities

You wake up to find your street has become a river, with water lapping at your doorstep. The sewers are overwhelmed, and the water continues to rise, seeping into your home and damaging your belongings. You're left to navigate the aftermath, dealing with cleanup, repairs, and the emotional toll of the event.

Understanding the common natural disasters that can impact your area is the first step in assessing your risks and vulnerabilities. By familiarizing yourself with these events and their potential consequences, you can better prepare yourself and your loved ones to face them head-on.

In the next section, we'll explore man-made emergencies and disasters, and how they differ from their natural counterparts. Stay tuned as we continue to build your knowledge base and empower you to become a master of disaster preparedness.

Man-Made Disasters and Emergencies

While natural disasters are caused by forces of nature, man-made disasters and emergencies arise from human activities, accidents, or malicious intent. These events can be just as devastating as their natural counterparts, and in some cases, even more complex to handle. Let's explore some common man-made disasters and emergencies:

a. Industrial Accidents
- Can involve explosions, fires, or the release of toxic substances
- Often occur due to equipment failure, human error, or safety violations
- Can have severe impacts on workers, nearby communities, and the environment
- May require specialized response teams and emergency management

Imagine you're living near an industrial park when suddenly, a deafening explosion rocks the area. A plume of toxic smoke rises from one of the factories, and sirens fill the air. You're faced with the decision to evacuate or shelter in place, uncertain of the extent of the danger.

b. Transportation Accidents
- Can involve cars, trucks, trains, planes, or ships
- May result in injuries, fatalities, and damage to infrastructure
- Can cause fires, explosions, or hazardous material spills
- Require coordinated emergency response and may disrupt travel and commerce

Picture yourself commuting to work when a massive pileup occurs on the highway ahead. Cars are crumpled, and debris litters the road. Emergency responders rush to the scene to aid the injured and clear the wreckage, leaving you stranded and wondering about the safety of your loved ones.

c. Hazardous Material Incidents
- Involve the release of toxic, flammable, or corrosive substances
- Can occur during transportation, storage, or use of these materials
- Pose risks to public health, safety, and the environment
- May require evacuation, specialized cleanup, and long-term monitoring

You're going about your day when a tanker truck overturns on a nearby road, spilling its contents into a local river. The water begins to change color, and a pungent odor fills the air. Authorities warn residents to avoid the area and seek medical attention if exposed to the hazardous substance.

d. Active Shooter Situations
- Involve one or more individuals using firearms to cause harm or death
- Can occur in public places, schools, workplaces, or other locations
- Require rapid response from law enforcement and emergency services
- Can have long-lasting emotional and psychological impacts on survivors and communities

Imagine you're at a shopping mall when the sound of gunfire erupts. Chaos ensues as people run for cover, unsure of where the threat is coming from. You're faced with the need to quickly assess the situation, find a safe location, and follow the instructions of first responders.

e. Cyber Attacks and Infrastructure Failures
- Can target computer systems, networks, or critical infrastructure
- May result in data breaches, financial losses, or disruptions to essential services
- Can have cascading effects on society, economy, and national security
- Require robust cybersecurity measures and resilience planning

You wake up to find that a widespread cyber attack has crippled the power grid, leaving your city without electricity. Cell phone networks are overloaded, and ATMs are offline. You're left to navigate the challenges of daily life without the technologies you've come to rely on.

Understanding the diverse range of man-made disasters and emergencies is crucial for comprehensive disaster preparedness. By recognizing the potential risks and impacts of these events, you can take steps to protect yourself, your family, and your community.

In the next section, we'll delve into the process of evaluating your local risks and vulnerabilities, helping you tailor your preparedness efforts to the specific challenges you may face. Get ready to assess your unique situation and build a resilient foundation for facing any crisis that comes your way.

Assessing Your Risks and Vulnerabilities

Now that you've gained an understanding of common natural disasters and man-made emergencies, it's time to assess your own risks and vulnerabilities. This process involves evaluating the specific hazards that could impact your area, as well as considering your personal circumstances and how they might affect your ability to prepare for and respond to a crisis.

Let's break down the key steps in assessing your risks and vulnerabilities:

a. Identify Local Hazards
- Research the natural disasters and man-made emergencies that have occurred in your area historically
- Consult local government agencies, emergency management organizations, and community resources for information on potential risks
- Consider factors such as geography, climate, and proximity to industrial sites or critical infrastructure

Imagine you've just moved to a new city and want to ensure you're prepared for any potential disasters. You start by visiting your local emergency management office and asking about the most common hazards in the area. They provide you with a list of historical events and a map highlighting flood zones, earthquake fault lines, and other risk factors.

b. Evaluate Your Home and Neighborhood
- Assess the structural integrity and safety features of your home, such as reinforced foundations, storm shutters, or fire-resistant materials

- Identify potential hazards in your neighborhood, such as overhanging trees, power lines, or nearby water bodies
- Consider your proximity to emergency services, evacuation routes, and community support networks

You take a walk around your property, noting any potential vulnerabilities. You notice that a large tree branch overhangs your roof, posing a risk during high winds. You also realize that your home is located in a low-lying area, making it more susceptible to flooding. With this information, you can prioritize mitigation measures and plan accordingly.

c. Consider Personal Factors
- Take into account your family's unique needs, such as medical conditions, disabilities, or special requirements for children or pets
- Evaluate your financial preparedness, including emergency savings, insurance coverage, and critical document storage
- Assess your social support network and identify trusted individuals who can assist you during a crisis

You sit down with your family to discuss your preparedness plans. Your elderly mother requires daily medication, and your young children have specific dietary needs. You realize that you'll need to stock extra supplies and create a plan for managing these challenges during an emergency. You also reach out to neighbors and friends, establishing a support network you can rely on.

d. Conduct a Risk Assessment Matrix

- Create a table or grid that lists potential hazards along one axis and the likelihood and impact of each hazard along the other axis
- Assign ratings to each hazard based on the probability of occurrence and the severity of consequences
- Use the matrix to prioritize your preparedness efforts and allocate resources to the most significant risks

You create a risk assessment matrix for your household, listing hazards like hurricanes, house fires, and power outages. You assign each hazard a rating based on its likelihood and potential impact. This visual representation helps you focus your efforts on the most critical risks and ensures that you're allocating your time and resources effectively.

By taking the time to assess your risks and vulnerabilities, you're laying the foundation for effective disaster preparedness. Understanding your unique situation allows you to tailor your plans, gather the necessary supplies, and build the skills and knowledge needed to face any challenge that comes your way.

In the next part of this book, we'll dive into the essential survival skills that every prepared individual should master. Get ready to learn how to shelter, hydrate, feed, and protect yourself and your loved ones in the face of adversity. The journey to becoming a master of disaster preparedness continues!

PART II
ESSENTIAL SURVIVAL SKILLS
SHELTER: FINDING AND BUILDING SAFE SPACES

In the face of a disaster, having a safe and secure shelter is essential for protecting yourself and your loved ones from the elements, as well as from potential threats. Whether you're forced to evacuate or shelter in place, knowing how to find or create a suitable shelter can mean the difference between life and death.

Let's explore the key aspects of finding and building safe spaces:

a. Identifying Suitable Shelter Locations
- Look for natural features that can provide protection, such as caves, dense forests, or rock formations
- Consider the safety and stability of existing structures, like buildings, bridges, or tunnels
- Evaluate the proximity to resources, such as water, food, and fuel sources
- Assess the risk of potential hazards, such as flooding, landslides, or falling debris

Imagine you've been forced to evacuate due to a wildfire rapidly approaching your community. As you navigate the wilderness, you come across a rocky outcropping that provides a natural barrier against the flames and smoke. You quickly assess the stability of the rocks and determine that it's a suitable location to take shelter until the danger passes.

b. Building Temporary Shelters
- Learn how to construct basic shelters using natural materials, such as branches, leaves, and grasses
- Practice building lean-tos, debris huts, or other simple structures that can provide protection from the elements
- Consider the insulation and ventilation of your shelter to regulate temperature and prevent moisture buildup
- Ensure your shelter is sturdy enough to withstand strong winds, heavy rain, or other potential hazards

You find yourself stranded in the wilderness after a hiking trip gone wrong. As night approaches, you realize you need to construct a temporary shelter to stay warm and dry. Using fallen branches and leaves, you build a simple lean-to against a large tree trunk. You cover the structure with a tarp from your backpack to provide additional protection from the rain.

c. Fortifying Existing Structures
- Identify the strengths and weaknesses of existing buildings or structures that could serve as shelter
- Reinforce doors, windows, and other potential entry points to improve security and protection
- Use sandbags, plywood, or other materials to create barriers against flooding or debris
- Ensure proper ventilation and air filtration to maintain a safe and healthy indoor environment

You've decided to shelter in place during a hurricane, and you need to quickly fortify your home. You start by boarding up windows with plywood to prevent shattering from high winds.

You also place sandbags around the perimeter of your house to divert floodwater. Inside, you create a safe room in a windowless interior space, stocked with essential supplies.

d. Maintaining and Improving Shelter Conditions
- Regularly assess the condition of your shelter and make necessary repairs or improvements
- Ensure proper sanitation and waste management to prevent the spread of disease
- Establish a fire safety plan and maintain proper ventilation to avoid carbon monoxide poisoning
- Create a comfortable and organized living space to maintain physical and mental well-being

After several days of sheltering in place, you notice that the roof of your shelter has started to leak. You quickly gather materials to patch the hole and reinforce the structure. You also take the time to organize your living space, designating areas for sleeping, storage, and food preparation. By maintaining a clean and comfortable shelter, you boost your morale and increase your chances of long-term survival.

Understanding how to find and create safe spaces is a crucial survival skill that can help you weather any disaster. By mastering the art of shelter-building and fortification, you'll be prepared to protect yourself and your loved ones in even the most challenging circumstances.

In the next section, we'll dive into another critical survival skill: water. Learn how to source, purify, and store this essential resource to ensure your survival in any situation.

WATER: SOURCING, PURIFICATION, AND CONSERVATION

In any survival situation, access to clean drinking water is vital. Without proper hydration, your body will quickly deteriorate, leading to fatigue, cognitive impairment, and eventually, death. Therefore, it's essential to know how to find, purify, and conserve water to ensure your survival.

Let's dive into the key aspects of water sourcing, purification, and conservation:

a. Identifying Potential Water Sources
- Locate natural water sources such as rivers, streams, lakes, and springs
- Learn to identify signs of water in the environment, such as lush vegetation, animal tracks, or bird flight paths
- Consider alternative water sources like rainwater, dew, or plant transpiration
- Be aware of potential contamination risks associated with each water source

Imagine you're lost in a remote wilderness area, and your water supplies are running low. You scan your surroundings and notice a cluster of green vegetation in the distance. As you approach, you discover a small stream flowing through the foliage. You've successfully identified a potential water source that could sustain you until help arrives.

b. Water Purification Methods
- Boiling: Bring water to a rolling boil for at least one minute to kill harmful bacteria and parasites

- Filtration: Use a commercial water filter or create a makeshift filter using natural materials like sand, charcoal, and cloth
- Chemical Treatment: Use water purification tablets or drops containing iodine or chlorine to disinfect water
- Solar Disinfection: Fill clear plastic bottles with water and expose them to direct sunlight for several hours

You've collected water from a nearby lake, but you're unsure of its purity. To be safe, you decide to boil the water. You build a small fire and place a pot of lake water over the flames. Once the water reaches a rolling boil, you let it continue for a full minute before removing it from the heat. The boiling process has killed any harmful microorganisms, making the water safe to drink.

c. Water Conservation Techniques
- Minimize water loss through perspiration by staying cool and limiting physical exertion during hot periods
- Collect and store water whenever possible, using containers like water bottles, tarps, or natural depressions
- Practice efficient water usage by prioritizing drinking, cooking, and hygiene needs
- Implement moisture-saving techniques like covering soil with mulch or using a solar still to extract water from plants

You've successfully found and purified a water source, but you know that you need to make your water supply last. You fill all available containers with clean water and store them in a cool, shaded area.

You also create a simple moisture trap by digging a small hole, filling it with green vegetation, and covering it with a clear plastic sheet. As the sun heats the vegetation, water condenses on the underside of the plastic, which you can collect and drink.

d. Staying Hydrated in Survival Situations
- Prioritize drinking water over food consumption, as your body can survive longer without food than without water
- Drink small amounts of water frequently, rather than large quantities at once, to optimize absorption and prevent dehydration
- Monitor your urine color and output as an indicator of hydration status (clear or pale yellow urine is a sign of adequate hydration)
- Be aware of the signs and symptoms of dehydration, such as thirst, dry mouth, fatigue, and dizziness

You've been rationing your water supply, but the heat and physical exertion of survival have taken a toll on your body. You notice that your urine has become dark and scanty, and you're experiencing a persistent thirst. Recognizing these signs of dehydration, you prioritize drinking small amounts of water at regular intervals to gradually replenish your fluid levels and maintain proper hydration.

By mastering the skills of water sourcing, purification, and conservation, you'll be equipped to maintain proper hydration and increase your chances of survival in any situation. Remember, water is life – prioritize it in your survival planning and preparation.

In the next section, we'll explore the importance of fire in survival situations. Learn how to start, maintain, and use fire safely to provide warmth, cook food, purify water, and signal for rescue.

FIRE: STARTING, MAINTAINING, AND USING FIRE SAFELY

In a survival situation, fire is a versatile and essential tool. It provides warmth, light, and comfort, allowing you to cook food, purify water, and signal for rescue. However, starting and maintaining a fire can be challenging, especially in adverse conditions. It's crucial to master the skills of fire-making and to understand how to use fire safely and responsibly.

Let's explore the key aspects of starting, maintaining, and using fire safely:

a. Gathering Fire-Making Materials
- Collect tinder (easily combustible materials like dry grass, leaves, or bark shavings) to help start your fire
- Find kindling (small twigs and branches) to help build and sustain the fire
- Gather larger logs or branches for fuel once the fire is established
- Ensure all materials are as dry as possible to facilitate fire-starting

Imagine you're setting up camp in a damp, wooded area. You scout the surrounding forest floor for dry tinder and kindling. You locate a standing dead tree and carefully collect small twigs and bark shavings from its branches. You also gather larger, dry branches from the tree to use as fuel once your fire is burning steadily.

b. Fire-Starting Techniques
- Friction methods: Use tools like a bow drill, hand drill, or fire plough to generate heat and ignite tinder
- Sparking methods: Employ ferrocerium rods, flint and steel, or a magnifying lens to create sparks and ignite tinder
- Modern methods: Utilize matches, lighters, or chemical fire starters as reliable and efficient options
- Practice fire-starting techniques in various weather conditions to build proficiency

You've gathered your fire-making materials and are ready to start your fire. You decide to use a ferrocerium rod and striker, as they work well even in damp conditions. You create a small pile of tinder and position the rod near the base. Holding the striker at a 45-degree angle, you quickly and firmly scrape it along the rod, directing the sparks toward the tinder. After a few attempts, the tinder ignites, and you carefully nurture the flame.

c. Maintaining and Controlling Fire
- Build your fire gradually, starting with tinder and slowly adding kindling and larger fuel as the fire grows
- Arrange logs or branches in a structure that allows for proper air flow, such as a teepee or pyramid shape
- Monitor the fire regularly and add fuel as needed to maintain desired heat and light
- Keep the fire contained within a designated fire pit or ring to prevent spreading

With your tinder ignited, you begin adding small kindling to the flame, gradually increasing the size of the twigs and branches as the fire grows. You arrange the larger fuel in a teepee structure around the kindling, ensuring that there is ample space for air to circulate. As the fire burns, you monitor it attentively, adding more fuel as needed to maintain a steady, controlled burn.

d. Fire Safety and Responsibility
- Choose a safe location for your fire, away from low-hanging branches, dry grass, or other flammable materials
- Clear the area around your fire pit of debris and create a barrier to contain the fire
- Keep a supply of water, sand, or dirt nearby to extinguish the fire if needed
- Fully extinguish the fire before leaving the area or going to sleep, stirring the ashes and dousing with water until no embers remain

As you tend to your fire, you remain mindful of safety. You've selected a location on bare dirt, away from overhanging branches and dry brush. You've also cleared a wide perimeter around the fire pit and have a bucket of water nearby in case of emergency. When it's time to extinguish the fire, you spread the embers and ashes, dousing them thoroughly with water. You stir the ashes and apply more water until the fire is completely extinguished and no heat remains.

Mastering the art of fire-making and understanding fire safety are critical survival skills. By learning how to start, maintain, and use fire responsibly, you'll be able to harness its power for warmth, cooking, water purification, and signaling, while minimizing the risk of unintended spread or harm.

In the next section, we'll delve into the world of food in survival situations. Learn how to forage for edible plants, set traps and snares, and preserve food to ensure a sustainable food supply in any environment.

Food: Foraging, Hunting, and Food Preservation Techniques

In a survival situation, access to food is critical for maintaining energy, mental clarity, and overall health. While you can survive for weeks without food, a lack of proper nutrition will gradually weaken your body and mind, making it harder to face the challenges of survival. By learning how to forage, hunt, and preserve food, you'll be able to sustain yourself and your loved ones in the face of adversity.

Let's explore the key aspects of foraging, hunting, and food preservation:

a. Foraging for Edible Plants
- Learn to identify common edible plants in your area, such as berries, nuts, roots, and leafy greens
- Study the distinguishing characteristics of each plant, including leaf shape, color, and growth patterns
- Be aware of poisonous plants that may resemble edible varieties, and learn how to differentiate between them
- Follow the Universal Edibility Test when trying unfamiliar plants to minimize the risk of adverse reactions

Imagine you're hiking through a dense forest and spot a patch of wild berries. Before consuming them, you carefully inspect the plant, comparing its features to those of known edible berries in your guidebook. You notice that the leaves, stem, and fruit match the description of a safe, edible variety.

To further confirm, you perform the Universal Edibility Test, rubbing a small portion of the plant on your skin and lips to check for any adverse reactions before consuming a small quantity and monitoring your response.

b. Hunting and Trapping Techniques
- Learn to identify signs of animal activity, such as tracks, droppings, and nesting sites
- Master the art of setting snares and traps to capture small game like rabbits, squirrels, and birds
- Practice using primitive hunting tools like bows, spears, and slingshots to take down larger game
- Understand the habits and behaviors of local wildlife to increase your chances of a successful hunt

You've set up camp in a wooded area and notice a well-worn game trail nearby. Recognizing this as a sign of regular animal activity, you decide to set a simple snare trap. Using cordage from your survival kit, you create a noose and attach it to a sturdy branch or stake. You position the snare along the game trail, ensuring that it blends in with the surroundings. By understanding the movement patterns of local wildlife, you've increased your chances of capturing a small animal for sustenance.

c. Fishing and Aquatic Food Sources
- Learn to identify edible aquatic plants, such as cattails, water lilies, and seaweed
- Master fishing techniques using improvised hooks, lines, and nets made from natural materials or survival kit components

- Familiarize yourself with the behavior and habitats of local fish species to optimize your fishing efforts
- Understand the importance of proper fish cleaning and preparation to avoid foodborne illnesses

While foraging near a lake, you spot a school of fish swimming near the shore. Using your survival knowledge, you fashion a simple hook from a small twig and some cordage. You bait the hook with a small piece of worm and cast your line into the water. After several attempts, you feel a tug on the line and reel in a decent-sized fish. You carefully clean and gut the fish, cooking it thoroughly over your campfire to ensure it's safe to eat.

d. Food Preservation Methods
- Learn to dry and smoke meat, fish, and vegetables to extend their shelf life and prevent spoilage
- Use salt or sugar to cure meat and create jerky for long-term storage
- Understand the principles of fermentation to create preserves, pickles, and other shelf-stable foods
- Store preserved foods in cool, dry places, protected from moisture and pests

After a successful day of foraging and hunting, you find yourself with an abundance of food. To ensure that nothing goes to waste, you decide to preserve some of your harvest. You create a simple drying rack using branches and cordage, and slice your meat and vegetables into thin strips. You place the strips on the rack and position it near your campfire, allowing the heat and smoke to dry and preserve the food.

By mastering food preservation techniques, you've created a reliable food supply that can sustain you for days or even weeks.

By honing your skills in foraging, hunting, and food preservation, you'll be able to secure a sustainable food supply in any survival situation. Remember to always prioritize safety and conservation when harvesting wild foods, taking only what you need and minimizing your impact on the environment.

In the next section, we'll discuss the importance of first aid and medical care in survival situations. Learn how to assess and treat common injuries and illnesses, and how to create an emergency medical kit using both natural and manufactured resources.

First Aid and Medical Care in Emergencies

In a survival situation, the risk of injury and illness is significantly higher due to the challenging environment and limited access to medical resources. Knowing how to assess and treat common medical emergencies can mean the difference between life and death. By mastering basic first aid skills and understanding how to use natural remedies, you'll be better prepared to handle health crises in the field.

Let's explore the key aspects of first aid and medical care in emergencies:

a. Assessing and Prioritizing Injuries
- Learn to perform a rapid head-to-toe assessment to identify and prioritize injuries
- Use the ABCDE approach (Airway, Breathing, Circulation, Disability, Exposure) to evaluate a patient's condition
- Understand the difference between life-threatening and non-life-threatening injuries
- Know when to seek additional medical help or initiate evacuation procedures

Imagine you're in a remote wilderness area when your companion sustains a fall, resulting in multiple injuries. You quickly assess the situation using the ABCDE approach, checking for any obstructions to their airway, monitoring their breathing and circulation, and evaluating their level of consciousness. You identify a deep laceration on their leg as the most pressing concern and begin to address it while continuing to monitor their overall condition.

b. Treating Common Injuries and Illnesses
- Master the techniques for controlling bleeding, such as applying direct pressure, elevating the wound, and using tourniquets when necessary
- Learn to recognize and treat signs of shock, hypothermia, and heat exhaustion
- Understand how to immobilize fractures and sprains using splints and slings made from natural materials or survival kit components
- Familiarize yourself with the signs and symptoms of common illnesses, such as dehydration, food poisoning, and infections

With the bleeding under control, you turn your attention to your companion's other injuries. You notice that their ankle is swollen and painful to the touch, indicating a possible sprain or fracture. Using branches and cordage from your survival kit, you fashion a makeshift splint to immobilize the joint and reduce further injury. You also monitor your companion for signs of shock, keeping them warm and elevating their legs to promote circulation.

c. Natural Remedies and Medicinal Plants
- Learn to identify and use common medicinal plants, such as yarrow for wound healing, elderberry for immune support, and willow bark for pain relief
- Understand the principles of herbal medicine, including proper preparation and dosage
- Familiarize yourself with the potential risks and side effects of natural remedies, and know when to seek conventional medical treatment

- Incorporate medicinal plants into your emergency medical kit for added versatility

As you continue to monitor your companion's condition, you recall the medicinal plants you've learned about in your survival training. You locate a patch of yarrow growing nearby and carefully harvest a few leaves. You clean the leaves and apply them to your companion's wound, knowing that yarrow has natural antiseptic and healing properties. You also prepare a willow bark tea to help alleviate their pain and reduce inflammation.

d. Building an Emergency Medical Kit
- Assemble a well-stocked first aid kit that includes essential items such as bandages, gauze, antiseptic, and pain relievers
- Include any personal medications or medical supplies needed for pre-existing conditions
- Supplement your kit with natural remedies and tools, such as medicinal plants, honey, and activated charcoal
- Regularly check and replace expired or depleted items to ensure your kit remains functional

Before embarking on your wilderness adventure, you carefully assembled a comprehensive emergency medical kit. In addition to conventional first aid supplies, you included a selection of natural remedies, such as honey for wound care and activated charcoal for treating poisoning or digestive issues. As you treat your companion's injuries, you're grateful for the foresight and preparation that went into building your kit.

By developing a strong foundation in first aid and medical care, you'll be better equipped to handle the challenges of survival. Remember to prioritize prevention and self-care, as maintaining your own health and well-being is crucial in any emergency situation.

In the next section, we'll explore the importance of navigation and signaling in survival situations. Learn how to orient yourself using natural cues, navigate using maps and compasses, and signal for help using a variety of techniques and tools.

NAVIGATION AND SIGNALING TECHNIQUESM

In a survival situation, knowing how to navigate and signal for help can be the key to a successful rescue or to reaching safety. Whether you're lost in the wilderness, stranded in a remote location, or caught in the aftermath of a disaster, having a solid understanding of navigation and signaling techniques can greatly improve your chances of survival.

Let's explore the key aspects of navigation and signaling:

a. Orienting Yourself Using Natural Cues
- Learn to identify cardinal directions using the sun's position at different times of day
- Use the stars, such as the North Star (Polaris) in the Northern Hemisphere, to determine direction at night
- Understand how to read natural indicators, such as moss growth on trees or the direction of prevailing winds, to help orient yourself
- Familiarize yourself with the unique geographical features and landmarks of your area

Imagine you've been hiking in a remote area when you realize you've lost sight of the trail. As the sun begins to set, you recall your navigation training and look for natural cues to help orient yourself. You notice that the moss on the nearby trees is thicker on one side, indicating the northern exposure. Using this information, along with your knowledge of the sun's position at sunset, you're able to determine the cardinal directions and plan your route back to the trail.

b. Navigating with Maps and Compasses
- Learn how to read and interpret topographic maps, including contour lines, scale, and key symbols
- Understand how to use a compass to take bearings and plot courses
- Practice combining map and compass skills to navigate in various terrains and weather conditions
- Always carry a physical map and compass as a backup to electronic navigation devices

Before setting out on your wilderness adventure, you make sure to pack a detailed topographic map of the area and a reliable compass. As you hike, you regularly refer to the map to track your progress and identify nearby landmarks. When you find yourself unsure of your exact location, you use your compass to take a bearing and plot your position on the map. By combining these tools and skills, you're able to navigate confidently and avoid getting lost.

c. Signaling for Help
- Familiarize yourself with the international distress signals, such as the SOS pattern (three short, three long, three short) for visual and auditory signals
- Learn how to create high-visibility signals using natural materials, such as arranging rocks or logs in a large "X" or "SOS" pattern
- Understand how to use reflective surfaces, such as mirrors or metallic objects, to create flashes of light that can be seen from a distance
- Incorporate signaling devices, such as whistles, flares, or signal mirrors, into your survival kit

After sustaining an injury that has left you immobilized, you realize you need to signal for help. Using your survival knowledge, you locate an open area nearby and begin to create a large "SOS" pattern using branches and rocks. You also take out the signal mirror from your survival kit and practice aiming the reflected light toward the horizon, hoping to catch the attention of potential rescuers. By employing multiple signaling techniques, you increase your chances of being spotted and rescued.

d. Staying Found and Maintaining Communication
- Always inform others of your planned route, destination, and expected return time before venturing into the wilderness
- Carry a means of communication, such as a satellite phone, two-way radio, or personal locator beacon (PLB), when traveling in remote areas
- Regularly update your contacts or leave messages at predetermined check-in points to maintain communication
- If you become lost or stranded, stay in one place to conserve energy and increase your chances of being found

Before embarking on your solo backpacking trip, you make sure to leave a detailed itinerary with a trusted friend, including your planned campsites and expected check-in times. You also carry a satellite communication device that allows you to send pre-programmed messages to your contacts, letting them know you're safe and on schedule. By taking these precautions and maintaining regular communication, you reduce the risk of becoming truly lost or stranded in the wilderness.

By mastering navigation and signaling techniques, you'll be better prepared to handle the challenges of survival and increase your chances of a successful rescue or safe return. Remember to always prioritize preparation and practice, as honing these skills before an emergency can make all the difference when you need them most.

In the next section, we'll discuss the importance of self-defense and security measures in survival situations. Learn how to assess and mitigate potential threats, create secure shelters, and protect yourself and your resources using a variety of strategies and tools.

Self-Defense and Security Measures

In a survival situation, ensuring your personal safety and protecting your resources are paramount. Whether you're facing the aftermath of a natural disaster, navigating a civil unrest, or encountering wildlife in a remote wilderness, understanding self-defense and security measures can help you stay safe and secure.

Let's explore the key aspects of self-defense and security in survival situations:

a. Situational Awareness and Threat Assessment
- Develop a keen sense of situational awareness, constantly observing your surroundings and identifying potential threats
- Learn to recognize signs of dangerous wildlife, such as bear scat or big cat tracks, and understand how to avoid and respond to encounters
- Be aware of human threats, such as looters or hostile individuals, and trust your instincts when a situation feels unsafe
- Understand the importance of maintaining a low profile and avoiding unnecessary attention or confrontation

Imagine you're navigating through an urban area in the aftermath of a natural disaster. As you make your way through the streets, you maintain a heightened sense of situational awareness, scanning your surroundings for potential threats. You notice a group of individuals acting suspiciously and decide to alter your route to avoid a potential confrontation. By staying alert and trusting your instincts, you're able to navigate the situation safely and securely.

b. Physical Self-Defense Techniques
- Learn basic self-defense techniques, such as striking vulnerable areas, breaking holds, and creating distance from an attacker
- Practice situational drills to develop muscle memory and improve your reaction time in high-stress situations
- Understand the principles of de-escalation and know when to retreat or avoid confrontation altogether
- Consider incorporating self-defense tools, such as pepper spray or a personal alarm, into your survival kit

While hiking through a remote wilderness area, you encounter an aggressive individual who attempts to steal your backpack. Recalling your self-defense training, you use verbal commands and body language to try to de-escalate the situation. When the attacker becomes physical, you employ targeted strikes to create space and allow for an escape. By combining situational awareness with practical self-defense techniques, you're able to protect yourself and your resources.

c. Fortifying Your Shelter and Camp
- Choose a secure location for your shelter or camp, considering factors such as natural barriers, clear lines of sight, and proximity to resources
- Construct physical barriers, such as walls or fences, using natural materials or survival kit components to deter intruders and wildlife
- Establish a perimeter and early warning system using trip wires, bells, or other noise-making devices to alert you to potential threats
- Maintain a clean and organized camp to avoid attracting unwanted attention from wildlife or human threats

After setting up camp in a remote forest, you begin to fortify your shelter and surrounding area. You construct a small wall using fallen branches and logs, creating a physical barrier between your camp and potential threats. You also set up a simple tripwire system using cordage and empty cans, ensuring that you'll be alerted to any intruders. By taking these proactive security measures, you create a safer and more defensible space for yourself.

d. Psychological and Emotional Resilience
- Understand the psychological impact of survival situations and develop strategies for managing stress, fear, and anxiety
- Practice positive self-talk and visualization techniques to maintain a resilient mindset in the face of adversity
- Foster a sense of community and cooperation with fellow survivors to create a support network and improve overall security
- Recognize the signs of post-traumatic stress and seek professional help when needed to address the long-term emotional impact of survival situations

As the days turn into weeks in your survival situation, you begin to feel the psychological toll of the experience. To maintain your emotional resilience, you practice daily mindfulness exercises, focusing on your breathing and visualizing positive outcomes. You also make an effort to connect with other survivors in your area, sharing resources and strategies for mutual support and security. By prioritizing your mental health and building a strong support network, you're better equipped to handle the ongoing challenges of survival.

By incorporating self-defense and security measures into your survival plan, you'll be better prepared to protect yourself, your loved ones, and your resources in the face of adversity. Remember that prevention and preparation are key, and that developing a robust skill set and mindset can make all the difference in a survival situation.

In the next part of this book, we'll delve into the process of emergency preparedness kit assembly, exploring the essential items and considerations for creating a well-rounded and personalized survival kit. Whether you're preparing for a short-term emergency or a long-term survival scenario, having the right tools and supplies can greatly enhance your chances of success.

PART III
EMERGENCY PREPAREDNESS KIT ASSEMBLY

THE IMPORTANCE OF AN EMERGENCY KIT

In any survival situation, having a well-stocked and carefully curated emergency kit can be the difference between life and death. An emergency kit, also known as a "bug-out bag" or "go-bag," is a collection of essential items and supplies that can help you survive for a period of time in the face of adversity. Whether you're facing a natural disaster, civil unrest, or a personal emergency, having a ready-to-go kit can provide you with the tools and resources you need to stay safe, healthy, and secure.

Let's explore the key reasons why an emergency kit is so crucial:

a. Immediate Access to Essential Supplies
- In the aftermath of a disaster or emergency, access to basic necessities like food, water, and medical supplies may be limited or cut off entirely
- By having a pre-packed emergency kit, you ensure that you have immediate access to these essential items when you need them most
- A well-stocked kit can sustain you and your loved ones for several days or even weeks, buying you valuable time to assess the situation and plan your next steps

Imagine a powerful earthquake strikes your city, causing widespread damage and disrupting essential services. With your emergency kit by your side, you have immediate access to clean drinking water, non-perishable food, first-aid supplies, and other critical items. While others may struggle to find these necessities in the chaos following the disaster, you and your family can focus on staying safe and planning your next moves.

b. Mobility and Flexibility in Uncertain Situations
- Emergencies and disasters can be unpredictable, and you may need to evacuate your home or navigate unfamiliar environments at a moment's notice
- An emergency kit is designed to be portable and easy to carry, allowing you to quickly grab it and go when the situation demands
- By having a mobile and self-contained kit, you maintain your flexibility and adaptability in the face of changing circumstances

Picture a wildfire rapidly approaching your community, forcing you to evacuate with little notice. With your emergency kit packed and ready to go, you can quickly load it into your vehicle and hit the road, knowing that you have the essential supplies you need to survive in the short term. As you navigate to a safer location, your kit provides you with the resources and peace of mind to adapt to the evolving situation.

c. Customization for Personal Needs and Circumstances
- Every individual and family has unique needs and considerations when it comes to emergency preparedness
- By assembling your own emergency kit, you can tailor the contents to your specific requirements, such as medical conditions, dietary restrictions, or regional hazards
- A personalized kit ensures that you have the items and supplies that are most relevant and useful to you, maximizing your chances of survival and comfort

Consider a family with a diabetic child and a elderly grandparent living in a hurricane-prone area. By creating a custom emergency kit, they can ensure that they have an adequate supply of insulin, syringes, and other medical necessities, as well as items like extra batteries for hearing aids and a backup supply of prescription medications. By tailoring their kit to their unique needs, they're better prepared to weather the challenges of a hurricane and its aftermath.

d. Peace of Mind and Empowerment
- The process of assembling an emergency kit can provide a sense of control and empowerment in the face of uncertain and potentially frightening situations
- By taking proactive steps to prepare for emergencies, you can reduce your anxiety and stress levels, knowing that you've done what you can to protect yourself and your loved ones
- An emergency kit represents a tangible investment in your own safety and well-being, providing a foundation of readiness and resilience

As you carefully select and pack the items for your emergency kit, you feel a growing sense of confidence and peace of mind. Each item represents a small but significant step towards being prepared for whatever challenges may come your way. As you zip up your completed kit, you feel empowered and ready to face the uncertainties of the future with a greater sense of control and security.

By understanding the vital role that an emergency kit plays in survival situations, you can approach the process of assembling your own kit with clarity, purpose, and confidence. In the following sections, we'll break down the essential components of a well-rounded emergency kit, providing you with the knowledge and guidance you need to create a personalized kit that meets your unique needs and circumstances.

Remember, an emergency kit is not just a collection of items - it's a lifeline, a symbol of your commitment to your own safety and well-being, and a powerful tool for navigating the challenges of survival.

WATER AND FOOD SUPPLIES FOR YOUR KIT

When assembling your emergency preparedness kit, water and food supplies should be among your top priorities. In the aftermath of a disaster or during a prolonged survival situation, access to clean drinking water and nutrient-rich food can be scarce. By including an adequate supply of these essentials in your kit, you can ensure that you and your loved ones have the sustenance you need to stay healthy and energized in the face of adversity.

Let's explore the key considerations for water and food supplies in your emergency kit:

a. Water: The Most Critical Resource
- The human body can only survive a few days without water, making it the most critical resource in your emergency kit
- Include at least one gallon of water per person per day, with a minimum three-day supply for evacuation scenarios and a two-week supply for home preparedness
- Store water in durable, leak-proof containers, such as BPA-free plastic bottles or collapsible water bags
- Consider including water purification methods, such as water filters, purification tablets, or bleach, to treat water from unknown sources

Imagine a severe storm has knocked out power and water services in your area. With your emergency kit's water supply, you have the peace of mind knowing that you and your family have access to clean, safe drinking water.

As you ration your water supply and use your purification tools to treat additional water from a nearby stream, you're able to stay hydrated and maintain your health during the crisis.

b. Non-Perishable, Nutrient-Dense Food
- Include a variety of non-perishable, nutrient-dense foods in your emergency kit to provide energy and essential nutrients
- Focus on items with long shelf lives, such as canned goods, dried fruits and nuts, energy bars, and dehydrated meals
- Consider any dietary restrictions or allergies within your household, and include appropriate food options
- Rotate your food supply regularly to ensure freshness and replace any expired items

Picture yourself and your family huddled in your home, riding out a severe winter storm that has left you without power for several days. As you open your emergency kit, you find a carefully curated selection of non-perishable food items, from hearty canned soups and stews to protein-rich nuts and dried meats. By having a diverse range of nutrient-dense foods at your fingertips, you're able to keep your energy levels up and maintain your overall health during the extended power outage.

c. Cooking and Preparation Tools
- Include necessary tools and utensils for preparing and consuming your emergency food supply

- Pack a manual can opener, eating utensils, and a small camping stove or heat source for cooking
- Consider including lightweight, reusable dishes and a small supply of dish soap for cleaning
- Remember to store any cooking fuels, such as propane or butane, in a separate, well-ventilated area

As you set up a makeshift camp after being forced to evacuate due to a wildfire, you rely on the cooking and preparation tools in your emergency kit to make the most of your food supply. Using your camping stove, you heat up a can of soup and boil some water for instant rice. With your reusable dishes and utensils, you're able to enjoy a hot, nourishing meal that helps to keep your spirits up and your body fueled for the challenges ahead.

d. Comfort Foods and Morale Boosters
- In addition to nutrient-dense staples, consider including a few comfort foods or morale boosters in your emergency kit
- Items like chocolate, hard candies, or individually-wrapped snacks can provide a much-needed mental and emotional lift during stressful times
- Be mindful of expiration dates and storage requirements for these items, and replace them as needed

After a long day of navigating the aftermath of an earthquake, you and your family take a moment to rest and recharge. As you dig through your emergency kit, you come across a small cache of your favorite candy bars and some packets of hot cocoa mix.

These small comfort foods provide a moment of normalcy and joy amidst the chaos, reminding you of the importance of tending to both your physical and emotional well-being in a crisis.

By thoughtfully selecting and packing water and food supplies for your emergency kit, you are taking a crucial step towards ensuring your survival and well-being in the face of adversity. Remember to regularly assess and update your supplies, taking into account changes in your household, dietary needs, and regional risks.

In the next section, we'll explore the essential first aid and medical supplies to include in your emergency kit, equipping you with the tools and knowledge to handle common injuries and health concerns in a survival situation.

First Aid and Medical Supplies

In any emergency or survival situation, the risk of injury or illness is significantly higher due to the challenging and often unpredictable environment. By including a well-stocked first aid kit and essential medical supplies in your emergency preparedness kit, you can ensure that you have the tools and resources needed to manage common injuries, treat minor illnesses, and provide life-saving care in the event of a more serious medical emergency.

Let's explore the key components of a comprehensive first aid and medical supply kit:

a. Basic First Aid Items
- Include a variety of bandages, gauze pads, and adhesive tape for wound care and bleeding control
- Pack antiseptic wipes, alcohol swabs, and antibiotic ointment to clean and protect wounds from infection
- Include a pair of scissors, tweezers, and safety pins for cutting bandages, removing splinters, and securing slings or splints
- Add a first aid manual or quick reference guide to help you properly assess and treat common injuries and illnesses

Imagine you're out on a hike when you accidentally slice your hand on a sharp rock. With your first aid kit on hand, you're able to quickly clean the wound with an antiseptic wipe, apply antibiotic ointment, and securely wrap the injury with a sterile bandage.

By having these basic first aid items readily available, you can prevent the wound from becoming infected and promote faster healing.

b. Medications and Treatments
- Include a supply of over-the-counter pain relievers, such as acetaminophen or ibuprofen, to manage pain and reduce inflammation
- Pack antihistamines and hydrocortisone cream to treat allergic reactions and insect bites
- Include antidiarrheal medication and oral rehydration salts to manage digestive issues and prevent dehydration
- Consider including prescription medications specific to your household's needs, such as inhalers for asthma or EpiPens for severe allergies

Picture a member of your family coming down with a high fever and severe headache in the midst of a power outage. With access to your emergency kit's supply of acetaminophen and a digital thermometer, you're able to monitor their temperature and provide relief from their symptoms until professional medical help becomes available.

c. Trauma and Emergency Care Supplies
- Include a tourniquet and hemostatic gauze for severe bleeding control
- Pack a SAM splint or moldable foam splints for immobilizing fractures and sprains
- Consider including a CPR mask or face shield for administering rescue breaths during CPR
- Add a space blanket or emergency bivvy to prevent shock and hypothermia in injured or ill individuals

During a severe earthquake, a heavy bookshelf falls on your partner, causing a deep laceration to their leg and significant bleeding. Using the tourniquet and hemostatic gauze from your emergency kit, you're able to stop the bleeding and stabilize their condition until emergency responders arrive. By having these advanced trauma care supplies on hand, you can provide potentially life-saving interventions in the critical moments following a serious injury.

d. Personal Protective Equipment (PPE)
- Include a supply of nitrile or latex gloves to protect yourself and others from bodily fluids and infections
- Pack a few N95 or surgical masks to reduce the risk of airborne illnesses and protect against dust and debris
- Consider including safety goggles or face shields to protect your eyes during first aid or rescue situations
- Add a bottle of hand sanitizer or disinfectant wipes to maintain proper hygiene and prevent the spread of germs

As you come across an injured neighbor in the aftermath of a hurricane, you take a moment to don your PPE from your emergency kit before providing assistance. By wearing gloves, a mask, and eye protection, you can safely assess their condition and provide care without putting yourself at risk of infection or exposure to hazardous materials.

By thoughtfully curating your first aid and medical supplies, you are equipping yourself with the tools and resources needed to handle a wide range of medical emergencies and health concerns in a survival situation.

Remember to regularly check and replace any expired or used items, and consider taking a first aid or CPR certification course to build your skills and confidence in providing emergency care.

In the next section, we'll discuss the importance of shelter and bedding essentials in your emergency preparedness kit, ensuring that you have the means to stay warm, dry, and protected in any environment.

Shelter and Bedding Essentials

In a survival situation, having access to proper shelter and bedding is essential for maintaining your physical health, mental well-being, and overall safety. Whether you're forced to evacuate your home or find yourself stranded in the wilderness, your emergency preparedness kit should include key items that can help you create a warm, dry, and secure shelter in any environment.

Let's explore the essential shelter and bedding components to include in your kit:

a. Shelter Options
- Include a high-quality, waterproof tent or tarp that can provide protection from the elements and create a makeshift shelter
- Consider adding a lightweight, durable sleeping bag or emergency bivvy that can keep you warm in cold temperatures
- Pack a set of sturdy, lightweight tent stakes and cordage for securing your shelter in windy conditions
- Include a compact, reflective emergency blanket that can be used for insulation, signaling, or as a ground cover

Imagine you're forced to evacuate your home due to a rapidly spreading wildfire. With your emergency kit in hand, you're able to quickly set up a sturdy tent in a safe location, providing a temporary shelter for you and your family. As night falls and temperatures drop, you're grateful for the insulated sleeping bags and emergency blankets that keep you warm and protected from the cold ground.

b. Insulation and Temperature Regulation
- Pack a set of thermal underwear or base layers that can wick away moisture and provide insulation in cold weather
- Include a few pairs of warm, moisture-wicking socks to keep your feet dry and prevent blisters and fungal growth
- Consider adding a lightweight, insulated jacket or vest that can be easily layered for added warmth
- Don't forget to include a warm hat, gloves, and a neck gaiter or scarf for protecting your extremities from the cold

After a severe blizzard leaves you stranded in your car on a remote stretch of highway, you're thankful for the insulation and temperature regulation items in your emergency kit. By layering your thermal underwear, warm socks, and insulated jacket, you're able to maintain your core body temperature and prevent hypothermia while you await rescue.

c. Bedding and Comfort Items
- Include a compact, inflatable sleeping pad or foam mat that can provide insulation and cushioning from the hard ground
- Pack a lightweight, packable pillow or stuff sack that can be filled with clothes to create a makeshift pillow
- Consider adding a few lightweight, quick-drying towels or washcloths for personal hygiene and comfort
- Don't forget to include a few personal comfort items, such as earplugs, an eye mask, or a small stuffed animal for children

As you set up camp in the wake of a devastating flood, you're relieved to have the bedding and comfort items from your emergency kit. By inflating your sleeping pad and using your packable pillow, you're able to create a relatively comfortable sleeping space that allows you to rest and recharge after a stressful day. The quick-drying towels and personal comfort items help to bring a sense of normalcy and hygiene to your temporary living situation.

d. Maintenance and Repair Supplies
- Include a small sewing kit with needles, thread, and scissors for repairing damaged clothing or gear
- Pack a few adhesive fabric patches or repair tape for fixing tears or holes in your tent or sleeping bag
- Consider adding a multitool or pocketknife that can be used for cutting cordage, making repairs, or constructing shelter
- Don't forget to include a few extra stakes, guylines, and bungee cords for reinforcing or modifying your shelter as needed

During an extended backcountry emergency, you notice a small tear in your tent's rainfly. Using the sewing kit and repair tape from your emergency kit, you're able to quickly patch the damage and prevent further water intrusion. By having these essential maintenance and repair supplies on hand, you can keep your shelter and bedding functional and effective, even in challenging conditions.

By carefully selecting and packing shelter and bedding essentials in your emergency preparedness kit, you are ensuring that you have the means to create a safe, warm, and comfortable living space in any survival situation. Remember to regularly inspect and maintain your gear, replacing any worn or damaged items as needed.

In the next section, we'll explore the importance of communication and lighting devices in your emergency kit, providing you with the tools to stay informed, signal for help, and navigate in low-light conditions.

COMMUNICATION AND LIGHTING DEVICES

In any emergency or survival situation, the ability to communicate with others and navigate in low-light conditions can be critical to your safety and well-being. By including a variety of communication and lighting devices in your emergency preparedness kit, you can ensure that you have the means to stay informed, signal for help, and maintain situational awareness in any environment.

Let's explore the key communication and lighting devices to include in your kit:

a. Two-Way Radios and Emergency Communication
- Include a set of high-quality, long-range two-way radios or walkie-talkies that can be used to communicate with family members or emergency responders
- Consider adding a hand-crank or battery-powered emergency radio that can receive NOAA weather alerts and local emergency broadcasts
- Pack a whistle or air horn that can be used to signal for help or attract attention in case of injury or distress
- Don't forget to include a laminated list of important emergency contact numbers and communication protocols

Imagine you're hiking in a remote area when your partner becomes injured and unable to walk. Using the two-way radios from your emergency kit, you're able to contact a nearby ranger station and request assistance. As you wait for help to arrive, you monitor the weather conditions and local emergency broadcasts on your hand-crank radio, ensuring that you stay informed and prepared for any changes in the situation.

b. Lighting Options
- Pack a high-quality, reliable headlamp or flashlight that can provide hands-free illumination in low-light conditions
- Include a set of long-lasting, high-performance batteries and a backup set of rechargeable batteries with a solar charge
- Consider adding a few glow sticks or chemical light markers that can be used for signaling or marking paths in the dark
- Don't forget to include a few candles or a compact lantern that can provide ambient light in a shelter or tent

During a power outage caused by a severe thunderstorm, you rely on the lighting options from your emergency kit to navigate your home safely. With your headlamp and flashlight, you're able to check on your family members, assess any damage to your property, and perform essential tasks like cooking or first aid. The glow sticks and candles help to provide a sense of comfort and normalcy in the darkness.

c. Backup Power and Charging
- Include a high-capacity, portable power bank or solar charger that can be used to charge cell phones, tablets, or other electronic devices
- Pack a few extra charging cables and adapters that are compatible with your devices and power sources
- Consider adding a compact, hand-crank or solar-powered generator that can be used to power small appliances or recharge batteries

- Don't forget to include a few spare batteries for your radios, headlamps, and other battery-operated devices

As you shelter in place during a hurricane, you use the backup power and charging devices from your emergency kit to keep your cell phone and tablet operational. With your portable power bank and solar charger, you're able to stay connected with loved ones, monitor the latest weather updates, and conserve your devices' battery life for essential communications.

d. Signaling and Visibility
- Pack a few high-visibility, reflective vests or armbands that can be used to increase your visibility to rescuers or emergency responders
- Include a set of flares, strobes, or signal mirrors that can be used to attract attention or guide rescuers to your location
- Consider adding a compact, waterproof strobe light or beacon that can be attached to your clothing or gear for hands-free signaling
- Don't forget to include a few brightly colored or reflective markers that can be used to mark trails or leave messages for rescuers

After becoming lost during a backcountry skiing trip, you use the signaling and visibility devices from your emergency kit to increase your chances of being found. By setting off a series of flares and using your signal mirror to reflect sunlight, you're able to attract the attention of a search and rescue helicopter. The high-visibility vest and strobe light help the rescuers to pinpoint your exact location and safely guide you to the helicopter.

By thoughtfully selecting and packing communication and lighting devices in your emergency preparedness kit, you are equipping yourself with the tools and resources needed to stay informed, signal for help, and navigate safely in any survival situation. Remember to regularly test and maintain your devices, replacing any expired batteries or damaged components as needed.

In the next section, we'll discuss the importance of hygiene and sanitation items in your emergency kit, ensuring that you have the means to maintain personal cleanliness and prevent the spread of disease in challenging circumstances.

Hygiene and Sanitation Items

Maintaining proper hygiene and sanitation is crucial in any emergency or survival situation, as it can help prevent the spread of disease, promote personal comfort, and boost morale. By including a variety of hygiene and sanitation items in your emergency preparedness kit, you can ensure that you have the means to stay clean, healthy, and comfortable in challenging circumstances.

Let's explore the essential hygiene and sanitation items to include in your kit:

a. Personal Hygiene Essentials
- Pack a supply of biodegradable soap, shampoo, and toothpaste that can be used for washing hands, bathing, and maintaining oral hygiene
- Include a set of compact, quick-drying towels or washcloths that can be used for drying off after washing or cleaning
- Consider adding a pack of disposable razors, a small mirror, and a set of nail clippers or scissors for personal grooming
- Don't forget to include a supply of feminine hygiene products, such as pads or tampons, for female members of your household

Imagine you've been evacuated to a temporary shelter after a major flood. With the personal hygiene essentials from your emergency kit, you're able to maintain a sense of cleanliness and normalcy despite the challenging circumstances.

By washing your hands regularly with biodegradable soap and keeping up with your oral hygiene routine, you can reduce your risk of illness and feel more comfortable in the crowded shelter environment.

b. Toilet and Waste Management
- Include a supply of toilet paper, tissues, or wet wipes that can be used for personal cleaning and waste management
- Pack a set of heavy-duty, biodegradable trash bags that can be used for collecting and disposing of waste
- Consider adding a compact, portable toilet or waste bag system that can be used in situations where regular toilets are unavailable
- Don't forget to include a small bottle of hand sanitizer or disinfectant wipes for cleaning hands after using the bathroom or handling waste

During a prolonged power outage, you find yourself without access to running water or flushing toilets. Using the toilet and waste management items from your emergency kit, you're able to set up a makeshift bathroom in your backyard. By using biodegradable trash bags and a portable waste bag system, you can safely and hygienically manage your waste until services are restored.

c. Laundry and Clothing Care
- Pack a small bottle of biodegradable laundry detergent or soap that can be used for washing clothes by hand
- Include a compact, portable clothesline or a few heavy-duty safety pins that can be used for hanging clothes to dry

- Consider adding a few mesh laundry bags or stuff sacks that can be used for organizing and washing smaller clothing items
- Don't forget to include a small sewing kit or a few patches and adhesive repair tape for fixing torn or damaged clothing

As you camp out in a remote area after a wildfire forces you to evacuate, you use the laundry and clothing care items from your emergency kit to keep your clothes clean and in good repair. By washing your clothes with biodegradable detergent and hanging them to dry on your portable clothesline, you can maintain a sense of hygiene and comfort in the outdoors.

d. Disease Prevention and First Aid
- Include a supply of disposable gloves, face masks, and eye protection that can be used for handling potentially contaminated materials or providing first aid
- Pack a bottle of disinfectant spray or wipes that can be used for cleaning surfaces, tools, or equipment that may be shared with others
- Consider adding a small bottle of insect repellent or a mosquito head net that can be used for preventing bites and reducing the risk of insect-borne illnesses
- Don't forget to include a supply of basic first aid items, such as bandages, antiseptic wipes, and over-the-counter medications for treating minor injuries or illnesses
- In the aftermath of a hurricane, you find yourself working alongside other volunteers to clean up debris and assist with recovery efforts. By using the disease prevention and first aid items from your emergency kit, you can protect yourself and others from potential health hazards.

With disposable gloves, face masks, and disinfectant wipes, you can safely handle contaminated materials and prevent the spread of germs, while your basic first aid supplies allow you to treat any minor injuries that may occur during the cleanup process.

By carefully selecting and packing hygiene and sanitation items in your emergency preparedness kit, you are ensuring that you have the means to maintain personal cleanliness, prevent the spread of disease, and promote overall health and well-being in any survival situation. Remember to regularly check and replace any expired or depleted items, and to adjust your supplies based on the specific needs and preferences of your household.

In the next section, we'll explore the importance of packing versatile tools and multipurpose gear in your emergency kit, providing you with the resources to tackle a wide range of challenges and tasks in a survival scenario.

Tools and Multipurpose Gear

In any emergency or survival situation, having access to a variety of tools and multipurpose gear can greatly enhance your ability to tackle challenges, solve problems, and improve your overall chances of success. By including a selection of versatile and reliable tools in your emergency preparedness kit, you can ensure that you have the means to build, repair, navigate, and adapt to a wide range of scenarios.

Let's explore the essential tools and multipurpose gear to include in your kit:

a. Multi-tools and Pocket Knives
- Include a high-quality, durable multi-tool that includes pliers, wire cutters, screwdrivers, and other useful functions
- Pack a reliable, sharp pocket knife or folding knife that can be used for cutting, whittling, or preparing food
- Consider adding a lightweight, compact multi-function shovel or trowel that can be used for digging, trenching, or clearing debris
- Don't forget to include a small sharpening stone or tool to keep your blades in top condition

Imagine you're setting up a makeshift shelter in the woods after a severe storm has left you stranded. With the multi-tools and pocket knives from your emergency kit, you're able to quickly and efficiently cut branches, lash together poles, and create a sturdy frame for your shelter. The compact shovel allows you to clear a level spot for your shelter and dig a small trench to divert rainwater.

b. Cordage and Repair Supplies
- Pack a supply of high-strength, lightweight cordage such as paracord or nylon rope that can be used for building shelters, securing gear, or creating snares
- Include a roll of duct tape or repair tape that can be used for fixing tears, sealing leaks, or creating makeshift splints
- Consider adding a few heavy-duty zip ties or bungee cords that can be used for securing or compressing items
- Don't forget to include a small sewing kit with needles, thread, and scissors for repairing clothes or gear

During a long-term power outage, you discover that a leaky window in your home is allowing cold air and moisture to enter. Using the cordage and repair supplies from your emergency kit, you're able to create a temporary seal around the window frame with duct tape and nylon rope. The zip ties and bungee cords help you secure a tarp over the exterior of the window to provide an extra layer of protection.

c. Navigation and Signaling Tools
- Include a high-quality, durable compass and a detailed topographic map of your local area
- Pack a reliable, waterproof GPS device or satellite communicator that can be used for navigation and emergency messaging
- Consider adding a compact, lightweight binocular or monocular for long-distance observation or signaling
- Don't forget to include a small mirror or whistle that can be used for signaling for help or attracting attention

While hiking in a remote wilderness area, you become disoriented and lose sight of the trail. Using the navigation tools from your emergency kit, you're able to orient yourself with your compass and map, identifying nearby landmarks and plotting a course back to safety. The GPS device provides additional confirmation of your location and allows you to send a message to a designated emergency contact, letting them know your situation.

d. Lighting and Fire-Starting Tools
- Pack a reliable, high-powered headlamp or flashlight with extra batteries
- Include a set of waterproof matches, a refillable lighter, or a ferrocerium rod for starting fires
- Consider adding a compact, folding saw or hatchet for processing firewood or clearing brush
- Don't forget to include a few chemical light sticks or a small lantern for area lighting

After a major earthquake disrupts power and gas lines in your area, you find yourself relying on the lighting and fire-starting tools from your emergency kit to provide warmth and illumination. With your headlamp and flashlight, you're able to safely navigate your home and check on your neighbors. The waterproof matches and ferrocerium rod allow you to start a small campfire in your backyard, providing a source of heat for cooking and boiling water.

By thoughtfully selecting and packing a range of tools and multipurpose gear in your emergency preparedness kit, you are equipping yourself with the resources and capabilities needed to adapt, improvise, and overcome challenges in any survival situation.

Remember to regularly inspect and maintain your tools, replacing any worn or damaged items as necessary.

In the next section, we'll discuss the importance of customizing your emergency kit to meet the specific needs and considerations of your household, ensuring that you have the right supplies and gear for your unique situation and environment.

Customizing Your Kit for Specific Needs

While a well-rounded emergency preparedness kit should include a core set of essential supplies and gear, it's equally important to customize your kit to meet the unique needs, preferences, and considerations of your household. By tailoring your kit to your specific situation, you can ensure that you have the right tools and resources to manage the challenges and risks that are most relevant to you and your loved ones.

Let's explore some key factors to consider when customizing your emergency kit:

a. Medical and Health Needs
- Include any prescription medications, medical supplies, or assistive devices that are regularly used by members of your household
- Pack extra glasses, contact lenses, or hearing aid batteries for those who rely on these items
- Consider adding specific items for managing chronic health conditions such as diabetes, asthma, or allergies
- Don't forget to include a copy of important medical information, such as prescription lists, doctor's contact information, and health insurance cards

Imagine you have a family member who relies on daily insulin injections to manage their diabetes. When customizing your emergency kit, you make sure to include a cooler with extra insulin, syringes, and glucose monitoring supplies. You also pack a few quick-acting carbohydrate snacks in case of low blood sugar episodes, and you include a laminated card with your family member's medical history and emergency contact information.

b. Dietary Restrictions and Preferences
- Pack emergency food supplies that align with any dietary restrictions or allergies in your household, such as gluten-free or nut-free options
- Include familiar and comforting food items that can provide a sense of normalcy and boost morale during stressful times
- Consider adding vitamins, supplements, or meal replacement bars to help meet nutritional needs
- Don't forget to include any necessary cooking tools or utensils, such as a manual can opener or a small camp stove

As you plan your emergency food supplies, you take into account your family's vegetarian lifestyle and your child's peanut allergy. You include a selection of plant-based protein sources, such as canned beans and textured vegetable protein, as well as nut-free energy bars and snacks. You also pack a few favorite comfort foods, like instant oatmeal packets and herbal tea bags, to provide a sense of familiarity and warmth during challenging times.

c. Climate and Environmental Factors
- Customize your clothing and bedding selections based on the typical weather patterns and temperature ranges in your area
- Include items that can provide protection from specific environmental hazards, such as dust masks for wildfire smoke or insect repellent for mosquito-prone regions
- Consider adding gear that can help you manage extreme temperatures, such as a portable fan for heat waves or extra insulation for cold snaps

- Don't forget to include supplies for managing specific risks in your area, such as sandbags for flooding or fire extinguishers for wildfire-prone regions

Living in a hurricane-prone coastal area, you make sure to customize your emergency kit with supplies that can help you manage the specific risks and challenges of tropical storms. You include a set of rain ponchos and waterproof boots for each family member, as well as a supply of plywood and tarps for covering windows and doors. You also pack a battery-powered radio and a waterproof phone case to help you stay informed and connected during the storm.

d. *Personal and Household Preferences*
- Include a few small comfort items or entertainment supplies that can provide a sense of normalcy and help pass the time during extended periods of sheltering in place
- Consider adding personal care items that align with your household's preferences, such as favorite brands of toothpaste or deodorant
- Pack a few items that can help you maintain important routines or traditions, such as a deck of cards for family game nights or a small journal for daily reflections
- Don't forget to include any necessary supplies for pets, such as food, leashes, or medications

As you customize your emergency kit, you think about the small things that can make a big difference in your family's comfort and well-being. You include a few paperback books and a set of coloring supplies to help keep your children entertained during long hours indoors.

You also pack a few scented candles and a small bottle of your favorite lotion to provide a sense of calm and normalcy amidst the chaos. For your dog, you include a collapsible water bowl and a few extra servings of their favorite kibble.

By taking the time to customize your emergency preparedness kit based on your household's specific needs, preferences, and considerations, you are creating a personalized resource that can provide both practical support and emotional comfort during challenging times. Remember to regularly reassess your kit and make updates as your circumstances or priorities change over time.

As you continue to build and refine your emergency preparedness plan, remember that your kit is just one part of a larger strategy for resilience and self-sufficiency. By combining your customized supplies with a strong foundation of knowledge, skills, and community connections, you can face any crisis with confidence and adaptability.

PART IV
DISASTER PLANNING AND PREPARATION
DEVELOPING A FAMILY EMERGENCY PLAN

Creating a comprehensive family emergency plan is a critical step in ensuring that you and your loved ones are prepared to face any crisis with clarity, coordination, and confidence. By taking the time to discuss potential risks, establish clear roles and responsibilities, and agree on key strategies and protocols, you can minimize chaos and confusion when disaster strikes and increase your chances of staying safe and connected.

Let's explore the essential components of a robust family emergency plan:

a. Identifying Risks and Scenarios
- Discuss the most likely and impactful emergency scenarios based on your location, climate, and community, such as natural disasters, power outages, or public health emergencies
- Consider potential secondary risks or complications that could arise from each scenario, such as road closures, supply chain disruptions, or communication breakdowns
- Identify any unique vulnerabilities or considerations within your household, such as medical needs, language barriers, or mobility limitations
- Don't forget to include often-overlooked risks, such as cybersecurity threats or financial emergencies

Imagine you live in an earthquake-prone area. As you develop your family emergency plan, you start by discussing the potential impacts of a major earthquake, such as structural damage to your home, injuries or entrapment, and prolonged power and water outages. You also consider secondary risks, like fires from ruptured gas lines or hazardous material spills. Additionally, you identify any specific vulnerabilities within your family, such as your elderly parent's limited mobility or your child's severe food allergies.

b. Establishing Roles and Responsibilities
- Assign clear roles and responsibilities to each family member based on their age, abilities, and experience, such as monitoring news updates, gathering supplies, or providing first aid
- Discuss backup plans and contingencies in case any family member is unable to fulfill their assigned role due to injury, absence, or other factors
- Consider creating a written or visual chart outlining each person's duties and how they contribute to the overall family response
- Don't forget to include any relevant contact information, such as work or school phone numbers, for each family member

As part of your family emergency plan, you assign specific roles to each household member. Your spouse, who has advanced first aid training, is designated as the primary medical responder. Your teenage daughter is tasked with monitoring official news sources and emergency alerts, while your son is responsible for gathering and organizing supplies from your emergency kit.

You create a clear visual chart outlining each person's duties and post it in a central location for easy reference.

c. Developing Communication and Reunion Plans
- Identify primary and secondary communication methods, such as phone trees, text messages, or designated out-of-area contacts, to ensure that all family members can stay informed and connected
- Establish a clear protocol for checking in and reporting status updates, including pre-determined time intervals and backup plans for missed check-ins
- Choose specific meeting places or reunion locations, both close to home and farther away, where family members can gather if separated or unable to communicate
- Don't forget to include important contact information, such as emergency services, utility companies, or insurance providers, in your communication plan

In your family emergency plan, you create a detailed communication and reunion strategy. You designate an out-of-state family member as your primary emergency contact, and all family members are instructed to send a brief status update to this contact as soon as possible after an emergency. You also establish a neighborhood meeting spot, such as a large tree in a nearby park, where family members should gather if unable to return home. Additionally, you identify a backup meeting location, like a relative's house in a neighboring town, in case your primary meeting spot is inaccessible.

d. Practicing and Maintaining Your Plan
- Schedule regular family meetings to review and update your emergency plan, ensuring that all members understand their roles and any changes to the strategy
- Conduct periodic drills or practice scenarios to test your plan's effectiveness and identify areas for improvement, such as evacuation routes or communication methods
- Keep physical and digital copies of your plan in easily accessible locations, such as your emergency kit, vehicle glove compartment, or cloud storage account
- Don't forget to share relevant parts of your plan with extended family, friends, or neighbors who may need to assist or coordinate with you during an emergency

To ensure that your family emergency plan remains effective and up-to-date, you schedule quarterly family meetings to review and discuss any necessary changes. You also conduct twice-yearly practice drills, such as a home fire evacuation or a family communication exercise, to test your plan's functionality and identify any gaps or weaknesses. Additionally, you keep a laminated copy of your plan in your emergency kit and share an electronic version with your designated out-of-area contact and close family friends.

By investing time and effort into developing a comprehensive family emergency plan, you are laying the foundation for a more coordinated, effective, and resilient response to any crisis that may come your way. Remember that your plan is a living document that should evolve and adapt as your family's circumstances change over time.

As you continue to refine your emergency preparedness strategy, consider how your family plan integrates with the broader disaster planning efforts in your community, workplace, and schools. By aligning your individual preparedness with the collective readiness of your networks and institutions, you can contribute to a more robust and interconnected safety net for all.

Establishing Communication Protocols

Effective communication is the backbone of any successful emergency response, allowing individuals and groups to share critical information, coordinate actions, and provide mutual support. By establishing clear and reliable communication protocols as part of your disaster planning process, you can ensure that you and your loved ones stay connected, informed, and empowered to make smart decisions under stress.

Let's explore the key steps in establishing robust communication protocols:

a. Identifying Primary and Secondary Methods
- Determine the most reliable and accessible communication methods for your family, considering factors such as coverage, cost, and ease of use
- Choose a primary method, such as cell phone calls or text messages, that will serve as your default mode of communication during an emergency
- Identify secondary or backup methods, such as email, social media, or satellite phones, that can be used if primary methods fail or become overloaded
- Don't forget to consider non-electronic methods, such as physical message boards or predetermined meeting places, as a last resort

Imagine a severe storm has caused widespread power outages and cell tower disruptions in your area. Thanks to your pre-established communication protocols, you know to switch from your primary method of cell phone calls to your secondary method of text messaging, which has a higher likelihood of getting through on overloaded networks. If text messages fail, you have a backup plan to leave a physical note at your designated neighborhood meeting spot, informing your family of your status and location.

b. Creating a Family Communication Plan
- Compile a comprehensive contact list including phone numbers, email addresses, and social media handles for all family members, as well as key contacts like doctors, insurance agents, and employers
- Designate an out-of-area contact, such as a friend or relative in another state, who can serve as a central point of communication and information sharing for your family
- Establish a clear protocol for checking in with each other and sharing status updates, including predetermined time intervals and prioritization of information
- Don't forget to include important medical information, such as allergies or prescription needs, in your communication plan for each family member

As part of your family communication plan, you create a laminated contact card for each family member to carry in their wallet or backpack. The card includes phone numbers and email addresses for all immediate family members, as well as your designated out-of-area contact.

You also establish a protocol where each family member will send a brief text message to the out-of-area contact every 4 hours during an emergency, providing a concise status update and location information. If anyone misses two consecutive check-ins, the out-of-area contact will attempt to reach them directly and share any known information with the rest of the family.

c. Familiarizing Yourself with Emergency Alert Systems
- Research the official emergency alert systems and notification methods used by your local government, utility companies, and weather service
- Sign up for any relevant opt-in alert programs, such as text message or email notifications, to ensure you receive timely and accurate information during a crisis
- Familiarize yourself with the meaning of different alert levels, such as "watch" vs. "warning," and the specific actions or precautions recommended for each level
- Don't forget to also monitor trusted news sources and social media accounts for breaking information and updates

In your area, the county government uses a combination of text message alerts, social media posts, and a dedicated emergency radio frequency to disseminate critical information during a disaster. You sign up for the county's opt-in text alert program and follow their official social media accounts on Facebook and Twitter. You also bookmark the website for your local news station, which has a reputation for providing timely and accurate coverage during emergencies. By familiarizing yourself with these alert systems and information sources ahead of time, you can quickly access and act upon crucial updates when every second counts.

d. Practicing and Updating Your Protocols
- Regularly review and update your communication protocols to ensure they remain relevant and effective, taking into account any changes in technology, family circumstances, or community alert systems
- Practice your communication protocols through periodic drills or simulations, such as sending test messages or conducting family check-in exercises
- Encourage family members to keep their contact information and communication devices up to date, and to always carry backup power sources like extra batteries or charging cables
- Don't forget to also share your communication protocols with extended family, friends, and neighbors who may need to reach you or assist in an emergency

To keep your communication protocols fresh and functional, you schedule a family meeting at the start of each year to review and update your contact lists, designated roles, and check-in procedures. You also conduct quarterly communication drills, where each family member practices sending a test status update to your out-of-area contact using various methods like text, email, and social media. Additionally, you make a habit of regularly checking and charging your backup communication devices, such as a spare cell phone battery or a hand-crank emergency radio, to ensure they are ready to use when needed.

By establishing clear, reliable, and well-practiced communication protocols, you are equipping yourself and your loved ones with the tools and strategies needed to stay connected and informed during even the most chaotic and stressful situations. Remember that effective communication is not just about having the right technology or information, but also about fostering trust, collaboration, and mutual support within your family and community.

As you refine your communication protocols, consider how they integrate with the broader disaster response plans and infrastructure in your area. By aligning your individual efforts with the collective resources and expertise of your local agencies and organizations, you can contribute to a more seamless and resilient emergency communication network for all.

Fortifying Your Home for Disasters

In the face of natural disasters and other emergencies, your home serves as your first line of defense and your primary shelter. By taking proactive steps to fortify your home against common hazards and vulnerabilities, you can significantly enhance your family's safety, security, and resilience during a crisis.

Let's explore some key strategies for fortifying your home against disasters:

a. Conducting a Home Hazard Assessment
- Identify the most likely and impactful disaster scenarios based on your home's location, climate, and surrounding environment, such as hurricanes, earthquakes, or wildfires
- Inspect your home's structural integrity, paying close attention to the foundation, roof, windows, and doors, and note any existing damage, weaknesses, or vulnerabilities
- Assess your home's utility systems, including electrical wiring, gas lines, and water pipes, and check for any potential safety risks or maintenance needs
- Don't forget to also consider non-structural hazards, such as unsecured furniture, hazardous materials, or flammable landscaping

Imagine you live in a hurricane-prone coastal area. As part of your home hazard assessment, you identify several key vulnerabilities, such as an aging roof with loose shingles, unprotected windows, and a large tree with overhanging branches near your house.

You also note that your home's electrical panel is located in a low-lying area prone to flooding, and that you have several large bookcases that could topple during strong winds. By systematically assessing your home's specific risks and weaknesses, you can prioritize your fortification efforts and resources for maximum impact.

b. Reinforcing Structural Elements
- Invest in structural upgrades and reinforcements based on your home's specific needs and vulnerabilities, such as hurricane shutters, earthquake retrofitting, or fire-resistant roofing materials
- Secure large appliances, such as refrigerators and water heaters, to wall studs or floors to prevent tipping or displacement during a disaster
- Install or upgrade window coverings, such as impact-resistant glass or metal shutters, to protect against shattering and flying debris
- Don't forget to also fortify your home's outdoor elements, such as anchoring free-standing structures like sheds or gazebos, and trimming or removing hazardous trees or branches

To address the vulnerabilities identified in your home hazard assessment, you take several key steps to reinforce your home's structural integrity. You hire a contractor to replace your aging roof with a new, hurricane-rated system featuring wind-resistant shingles and reinforced trusses. You also install metal storm shutters over all your windows and doors, and securely anchor your large bookcases and appliances to the walls and floors.

Additionally, you trim back the overhanging tree branches and remove a deteriorating shed from your backyard to reduce potential hazards. By investing in these targeted structural upgrades, you significantly enhance your home's ability to withstand the impacts of a major hurricane.

c. Enhancing Utility Safety and Backup Systems

- Install or upgrade safety devices on your home's utility systems, such as surge protectors for electrical appliances, automatic shut-off valves for gas lines, and backflow prevention devices for water pipes
- Invest in backup power sources, such as a portable generator or solar panels, to maintain essential functions like refrigeration, lighting, and communication during outages
- Develop a plan for safely storing and using hazardous materials, such as cleaning supplies or fuel, and ensure proper ventilation and fire prevention measures are in place
- Don't forget to also have your utility systems regularly inspected and maintained by qualified professionals to ensure optimal safety and performance

As part of your home fortification efforts, you take steps to enhance the safety and resilience of your utility systems. You install a whole-house surge protection system to safeguard your electrical appliances from power spikes, and you have a licensed plumber add a backflow prevention device to your main water line to prevent contamination during a flood. You also purchase a portable propane generator and develop a plan for safely storing and rotating your fuel supply.

Additionally, you schedule annual inspections of your gas furnace and water heater to ensure they are operating safely and efficiently. By proactively addressing potential utility risks and establishing backup systems, you minimize the likelihood and impact of dangerous malfunctions or prolonged outages during a disaster.

d. Developing a Home Maintenance and Preparedness Checklist

- Create a comprehensive checklist of recurring home maintenance tasks and preparedness measures, such as cleaning gutters, testing smoke alarms, and updating emergency kits
- Assign specific tasks and timelines to each family member, taking into account their age, abilities, and availability
- Keep physical and digital copies of your checklist in easily accessible locations, such as on your refrigerator or in a shared online document
- Don't forget to also include seasonal or event-specific tasks, such as winterizing pipes before a cold snap or securing outdoor furniture before a hurricane

To help ensure that your home fortification efforts remain consistent and effective over time, you develop a detailed home maintenance and preparedness checklist. Your checklist includes regular tasks like cleaning dryer vents, checking fire extinguishers, and restocking first-aid kits, as well as seasonal measures like caulking windows before winter and cleaning your chimney before using your fireplace.

You assign each task to a specific family member and establish a clear schedule for completion, with reminders set in your shared digital calendar. You also print out a copy of the checklist and post it on your kitchen bulletin board for easy reference. By institutionalizing home maintenance and preparedness as a routine family practice, you create a culture of ongoing vigilance and responsibility that can pay significant dividends when disaster strikes.

By taking a proactive and systematic approach to fortifying your home against disasters, you are investing in the long-term safety, security, and resilience of your family and your property. Remember that effective home fortification is not a one-time event, but an ongoing process of assessment, upgrades, and maintenance that requires sustained attention and resources.

As you continue to refine your home disaster preparedness plan, consider how your individual efforts fit into the broader context of your neighborhood and community. By coordinating with your neighbors, sharing resources and expertise, and advocating for collective preparedness measures, you can help create a more secure and mutually supportive environment for all.

Vehicle Preparedness and Maintenance

In many disaster scenarios, your vehicle serves as a critical lifeline, providing transportation, shelter, and storage for your family and your emergency supplies. However, vehicles are also vulnerable to a range of hazards and breakdowns that can compromise their reliability and safety during a crisis. By prioritizing vehicle preparedness and maintenance as part of your overall disaster planning, you can ensure that your car or truck is ready to perform when you need it most.

Let's explore some essential strategies for maintaining and preparing your vehicle for emergencies:

a. Performing Regular Maintenance and Inspections
- Adhere to your vehicle's recommended maintenance schedule for oil changes, tire rotations, and other routine services to keep your car in optimal operating condition
- Conduct regular visual inspections of your vehicle's key components, such as brakes, hoses, belts, and fluid levels, and address any signs of wear, leaks, or damage promptly
- Pay special attention to your vehicle's battery, testing its charge and replacing it as needed to avoid unexpected failures during a crisis
- Don't forget to also keep your vehicle's exterior clean and rust-free, as corrosion can weaken structural integrity and compromise safety

Imagine you're preparing for the upcoming winter storm season. As part of your vehicle preparedness checklist, you take your car in for a full tune-up, including an oil change, brake inspection, and tire rotation.

You also visually inspect your vehicle's hoses and belts for cracks or fraying, and you test your battery's charge using a multimeter. Additionally, you wash and wax your vehicle's exterior, paying close attention to any areas of paint damage or rust. By proactively maintaining your vehicle's key systems and components, you reduce the likelihood of a breakdown or malfunction during a critical moment.

b. Equipping Your Vehicle with Emergency Supplies
- Create a vehicle-specific emergency kit that includes essential items like jumper cables, a tire inflator, a spare tire, and a jack
- Pack a set of basic tools, such as a wrench, pliers, and a screwdriver, as well as duct tape, zip ties, and a multi-tool for quick repairs
- Include a supply of non-perishable food, water, and warm clothing in case you become stranded or need to shelter in your vehicle for an extended period
- Don't forget to also pack a backup phone charger, a flashlight with extra batteries, and a physical map of your area in case electronic navigation fails

As part of your vehicle preparedness efforts, you assemble a dedicated emergency kit for your car. In addition to the standard items like jumper cables and a spare tire, you include a set of basic hand tools, several quarts of motor oil, and a pack of emergency flares. You also store a backpack containing non-perishable snacks, bottled water, a warm blanket, and a change of clothes in your trunk. Additionally, you keep a laminated list of important phone numbers and a regional map in your glove compartment.

By equipping your vehicle with a comprehensive set of emergency supplies, you enhance your ability to handle both routine breakdowns and more serious crisis situations.

c. Planning and Practicing Emergency Vehicle Procedures
- Develop a set of clear, step-by-step procedures for responding to common vehicle emergencies, such as a flat tire, an overheated engine, or a dead battery
- Practice these procedures with all family members who may need to drive the vehicle, ensuring that everyone knows how to safely operate the jack, change a tire, or jump-start the battery
- Establish a protocol for communication and reunification if family members become separated while traveling in different vehicles
- Don't forget to also plan and practice vehicle evacuation procedures, including identifying multiple routes to your designated meeting place and handling heavy traffic or road closures

To reinforce your vehicle emergency preparedness, you organize a family training day focused on practicing key procedures. You demonstrate how to safely jack up the car and change a tire, and you have each family member take turns locating and connecting the jumper cables. You also review the steps for handling an overheated engine, emphasizing the importance of allowing the engine to cool before attempting to open the radiator cap. Additionally, you discuss your family's communication and reunification plan, ensuring that everyone knows how to contact your designated out-of-area relative and navigate to your agreed-upon meeting place.

By actively practicing and discussing these procedures, you build your family's confidence and competence in handling vehicle-related emergencies.

d. Maintaining Fuel Levels and Alternate Transportation Options

- Make a habit of keeping your vehicle's gas tank at least half full at all times, as gas stations may be closed or overwhelmed during a disaster
- Consider investing in a portable gas can and storing a backup supply of fuel in a cool, dry place away from your home
- Research alternate transportation options in your area, such as public transit routes, bike paths, or ride-sharing programs, in case your primary vehicle is unavailable or impractical
- Don't forget to also maintain any secondary vehicles or equipment, such as motorcycles, bicycles, or trailers, that could serve as backup transportation in an emergency

As part of your vehicle preparedness routine, you make a point of filling up your gas tank whenever it drops below half full. You also purchase a five-gallon gas can and store it in your detached garage, rotating the fuel every six months to ensure freshness. In addition to your primary car, you regularly maintain your family's bicycles, ensuring that the tires are properly inflated and the brakes are in good working order. You also research the public bus and light rail routes in your area, identifying which lines run closest to your home and workplace. By maintaining fuel levels and exploring alternate transportation options, you increase your flexibility and resilience in navigating the challenges of a disaster.

By taking a comprehensive and proactive approach to vehicle preparedness and maintenance, you are not only investing in the reliability and safety of your transportation but also in the overall resilience and adaptability of your family in the face of emergencies. Remember that effective vehicle preparedness is an ongoing process that requires regular attention, practice, and resources.

As you refine your vehicle preparedness plan, consider how it integrates with your broader disaster planning efforts, such as your communication protocols, your home fortification measures, and your community's emergency response infrastructure. By creating a holistic and interconnected preparedness strategy, you can optimize your family's ability to anticipate, respond to, and recover from a wide range of potential crisis scenarios.

Financial Readiness and Document Protection

Disasters not only pose threats to our physical safety and property but can also have significant and long-lasting impacts on our financial well-being and legal identities. From the immediate costs of evacuation and sheltering to the long-term challenges of rebuilding and recovering, financial preparedness plays a critical role in a family's overall resilience. Additionally, protecting important documents and records is essential for preserving your legal rights, accessing benefits, and avoiding identity theft in the aftermath of a disaster.

Let's explore some key strategies for enhancing your financial readiness and document protection:

a. Building an Emergency Fund
- Set a goal to save at least three to six months' worth of living expenses in a dedicated emergency fund to cover unexpected costs during a disaster
- Keep your emergency fund in a secure, easily accessible account, such as a high-yield savings account or a short-term certificate of deposit
- Regularly contribute to your emergency fund, treating it as a non-negotiable part of your monthly budget
- Don't forget to also have a small amount of cash on hand in case ATMs or electronic payment systems are down during a crisis

Imagine a wildfire forces you to evacuate your home with little notice. Thanks to your dedicated emergency fund, you have the financial flexibility to cover the costs of temporary lodging, food, and other essentials without relying on credit cards or loans. You also have a few hundred dollars in cash stored in your evacuation bag, allowing you to make necessary purchases even if power outages or network disruptions prevent the use of debit or credit cards. By building a robust emergency fund and maintaining a cash reserve, you create a financial safety net that can help cushion the blow of unexpected expenses during a disaster.

b. Organizing and Protecting Important Documents
- Create a master inventory of your important documents, such as birth certificates, social security cards, insurance policies, and property deeds
- Store physical copies of these documents in a fireproof and waterproof safe or a secure off-site location, such as a safe deposit box
- Digitize your important documents by scanning them and storing them on a secure cloud-based platform or an encrypted external hard drive
- Don't forget to also have copies of your important documents in your evacuation bag or stored with a trusted out-of-area contact

As part of your document protection efforts, you purchase a small fireproof and waterproof safe to store your family's birth certificates, passports, and other critical records. You also spend a weekend scanning all your important documents and uploading them to a secure, encrypted cloud storage service.

Additionally, you make physical copies of your health insurance cards, driver's licenses, and credit cards, and you store them in a sealed envelope in your evacuation bag. By organizing and protecting your important documents, you ensure that you have the necessary information and proof to access services, file claims, and rebuild your life after a disaster.

c. Reviewing and Updating Insurance Coverage
- Review your homeowners or renters insurance policy to ensure that you have adequate coverage for the specific risks in your area, such as floods, earthquakes, or wildfires
- Consider purchasing additional riders or standalone policies for high-value items, such as jewelry, art, or electronics, that may have limited coverage under your standard policy
- Maintain an up-to-date inventory of your possessions, including photos and receipts, to streamline the claims process in case of damage or loss
- Don't forget to also review and update your health, life, and auto insurance policies to ensure that they meet your family's current needs and reflect any changes in your circumstances

To bolster your financial readiness, you schedule an annual review of your insurance policies with your agent. During the review, you discover that your homeowners policy has a high deductible for hurricane damage, so you decide to increase your coverage and add a separate flood insurance policy. You also create a detailed inventory of your possessions, complete with photos and purchase receipts, and you store a copy of the inventory in your cloud storage account.

Additionally, you update the beneficiaries on your life insurance policy to reflect the birth of your new child. By regularly reviewing and updating your insurance coverage, you can minimize gaps and ensure that you have the financial protection you need to recover from a disaster

d. Establishing Emergency Credit and Communication Plans

- Research and pre-qualify for emergency credit options, such as a home equity line of credit or a low-interest personal loan, that you can tap for immediate cash needs during a disaster
- Keep a list of your bank and credit card account numbers, as well as the contact information for your financial institutions, in a secure location that you can access during an emergency
- Establish a plan for communicating with your financial institutions and monitoring your accounts for fraudulent activity in the aftermath of a disaster
- Don't forget to also have a plan for paying your bills and managing your finances if you're displaced or unable to access your usual payment methods

As part of your financial readiness plan, you apply for a small personal line of credit from your local bank, which you can use for emergency expenses without having to liquidate your savings or investments. You also make a list of all your bank and credit card account numbers and customer service phone numbers, and you store a copy of the list in your password-protected cloud storage account. Additionally, you set up online bill pay for your mortgage, utilities, and other regular expenses, ensuring that you can manage your payments even if you're unable to mail checks or visit a branch in person.

By establishing emergency credit and communication plans, you create additional layers of financial flexibility and security to help you weather the challenges of a disaster.

By prioritizing financial readiness and document protection as part of your overall disaster planning, you are taking proactive steps to safeguard your economic well-being and preserve your legal identity in the face of unexpected challenges. Remember that effective financial preparedness is an ongoing process that requires regular attention, review, and adjustment to reflect changes in your circumstances and the evolving risks in your environment.

As you refine your financial readiness and document protection strategies, consider how they integrate with your broader disaster planning efforts, such as your emergency communication protocols, your evacuation plans, and your post-disaster recovery strategies. By creating a holistic and interconnected preparedness framework, you can maximize your family's ability to anticipate, withstand, and bounce back from the financial shocks and legal disruptions that can accompany a major disaster event.

Psychological Preparation and Mental Resilience

Disasters and emergencies can take a significant toll on our mental health and emotional well-being, often triggering feelings of fear, anxiety, grief, and helplessness. The stress of facing life-threatening situations, the trauma of loss and displacement, and the challenges of adapting to a radically altered environment can strain even the most resilient individuals and families. By prioritizing psychological preparation and mental resilience as part of your overall disaster planning, you can develop the skills, strategies, and mindset needed to cope with adversity, maintain hope, and support your loved ones through difficult times.

Let's explore some key strategies for cultivating psychological preparedness and building mental resilience:

a. Developing a Positive and Proactive Mindset
- Cultivate a sense of realistic optimism, focusing on your strengths, resources, and abilities to overcome challenges and find solutions
- Reframe potential disasters as opportunities for growth, learning, and building resilience, rather than as insurmountable threats or inevitable catastrophes
- Practice proactive problem-solving and decision-making, breaking down complex challenges into manageable steps and taking concrete actions to prepare and protect yourself and your loved ones
- Don't forget to also nurture a sense of humor and creativity, finding moments of lightness and inspiration even in the midst of difficult circumstances

Imagine you're facing the prospect of a major hurricane hitting your area. Rather than getting caught up in worst-case scenario thinking or media-fueled panic, you choose to focus on the concrete steps you can take to prepare and protect your family. You create a detailed emergency plan, gather supplies, and communicate with your loved ones, channeling your energy into proactive problem-solving. You also make a point of finding small moments of joy and connection with your family, such as playing board games or sharing stories, to maintain a sense of perspective and resilience. By cultivating a positive and proactive mindset, you build your psychological resources and increase your ability to navigate the challenges ahead.

b. Building a Strong Support Network

- Cultivate close, supportive relationships with family, friends, and neighbors, creating a network of people you can turn to for practical and emotional support during a crisis
- Participate in community preparedness efforts, such as local emergency response teams or neighborhood watch programs, to build social connections and a sense of collective resilience
- Identify and connect with mental health professionals, support groups, or crisis hotlines that you can access for additional support and guidance if needed
- Don't forget to also practice self-care and set healthy boundaries, recognizing that you need to prioritize your own well-being in order to effectively support others

As part of your psychological preparedness efforts, you make a point of strengthening your social connections and building a robust support network. You organize a neighborhood preparedness meeting, where you and your neighbors share ideas, resources, and contact information. You also join a local Community Emergency Response Team (CERT) program, where you receive training in disaster response skills and build relationships with other engaged citizens. Additionally, you research mental health resources in your area, such as counseling services and support groups, and you save the contact information for a crisis hotline in your phone. By actively cultivating a strong support network, you create a web of social and emotional resources that can help sustain you through the challenges of a disaster.

c. Practicing Stress Management and Coping Skills
- Learn and practice relaxation techniques, such as deep breathing, progressive muscle relaxation, or mindfulness meditation, to help manage stress and anxiety
- Develop a toolkit of healthy coping strategies, such as exercising, journaling, or engaging in creative activities, that you can turn to during times of stress or uncertainty
- Prioritize self-care activities, such as getting adequate sleep, eating nutritious meals, and staying hydrated, to maintain your physical and emotional well-being
- Don't forget to also practice self-compassion and avoid unhealthy coping mechanisms, such as excessive alcohol use or isolation, that can undermine your resilience

To enhance your mental resilience, you commit to a daily practice of stress management and self-care. You start each morning with a few minutes of deep breathing and gentle stretching, setting a calm and centered tone for the day. You also make a list of healthy coping strategies, such as going for a walk in nature, listening to music, or calling a friend, and you make a point of incorporating these activities into your routine, especially during times of high stress. Additionally, you prioritize getting at least seven hours of sleep each night and eating a balanced diet rich in fruits, vegetables, and whole grains. By proactively managing your stress and cultivating healthy habits, you build your psychological reserves and increase your ability to bounce back from adversity.

d. Fostering a Sense of Purpose and Meaning
- Reflect on your core values, beliefs, and goals, and consider how they can provide a sense of direction and motivation during challenging times
- Look for opportunities to make a positive difference in your community, such as volunteering, donating resources, or supporting others in need, to cultivate a sense of purpose and connection
- Practice gratitude and perspective-taking, regularly acknowledging the good things in your life and the lessons you can learn from difficult experiences
- Don't forget to also celebrate your successes and milestones, no matter how small, to maintain a sense of progress and accomplishment

As part of your psychological preparedness journey, you take time to reflect on what matters most to you and what gives your life a sense of meaning and purpose. You identify your core values, such as family, compassion, and resilience, and you consider how you can live out these values in your daily life, even in the midst of a crisis. You also look for ways to make a positive difference in your community, such as volunteering at a local food bank or checking in on elderly neighbors. Additionally, you start a gratitude journal, where you write down three things you're thankful for each day, helping to shift your focus from stress and worry to appreciation and perspective. By actively cultivating a sense of purpose and meaning, you tap into a powerful source of psychological strength and motivation that can help sustain you through even the toughest challenges.

By prioritizing psychological preparation and mental resilience as part of your overall disaster planning, you are developing the inner resources and strategies needed to face adversity with courage, compassion, and hope. Remember that building mental resilience is an ongoing process that requires regular practice, reflection, and support from others.

As you refine your psychological preparedness plan, consider how it integrates with your broader disaster planning efforts, such as your communication protocols, your self-care routines, and your community engagement strategies. By creating a holistic and proactive approach to resilience, you can enhance your ability to adapt, recover, and thrive in the face of life's inevitable challenges and disruptions.

Community Involvement and Resource Sharing

Disasters and emergencies are inherently social events, affecting not just individuals and families but entire communities and regions. While personal preparedness is essential, true resilience lies in the collective strength, resources, and support of interconnected networks of neighbors, organizations, and institutions. By prioritizing community involvement and resource sharing as part of your overall disaster planning, you can contribute to a more cohesive, adaptable, and empowered social fabric that can withstand and recover from even the most severe challenges.

Let's explore some key strategies for fostering community involvement and resource sharing:

a. Participating in Community Preparedness Efforts
- Attend local emergency preparedness meetings, workshops, and training events to learn about your community's risks, resources, and response plans
- Join or support community-based organizations, such as Community Emergency Response Teams (CERT), Neighborhood Watch programs, or faith-based groups, that focus on disaster preparedness and resilience
- Volunteer your time, skills, and resources to help develop and implement community preparedness projects, such as creating a neighborhood emergency communication plan or organizing a community resource fair
- Don't forget to also share your own knowledge, experiences, and perspectives with others, contributing to a collective body of wisdom and best practices

Imagine your local fire department is hosting a series of community preparedness workshops. You attend the workshops and learn about the specific risks and hazards in your area, as well as the emergency response protocols and resources available. Inspired by what you've learned, you decide to join your neighborhood's CERT program, where you receive training in basic disaster response skills, such as first aid, search and rescue, and fire safety. As part of the CERT team, you help organize a community preparedness fair, where local organizations and businesses come together to share resources, information, and support with residents. By actively participating in community preparedness efforts, you not only enhance your own knowledge and skills but also contribute to a more informed, connected, and resilient social ecosystem.

b. Building Relationships and Networks
- Get to know your neighbors, including their names, contact information, and any special skills, resources, or needs they may have in an emergency
- Participate in community events, such as block parties, neighborhood cleanups, or local festivals, to build social connections and a sense of shared identity
- Join or create local groups or online forums focused on disaster preparedness, where you can share information, ideas, and support with others in your community
- Don't forget to also cultivate relationships with local leaders, emergency responders, and media outlets, who can be valuable allies and resources during a crisis

As part of your community involvement efforts, you make a point of reaching out to your neighbors and building positive relationships. You organize a neighborhood potluck, where everyone brings a dish to share and takes turns introducing themselves and sharing a bit about their background and interests. You also create a neighborhood directory, with contact information and emergency details for each household, and distribute copies to everyone on your block. Additionally, you join a local Facebook group focused on disaster preparedness, where you can ask questions, share resources, and connect with other engaged citizens in your area. By proactively building relationships and networks, you create a web of social capital and support that can be activated and leveraged during times of crisis.

c. Establishing Resource-Sharing Agreements
- Work with your neighbors, community organizations, and local businesses to identify and map the skills, resources, and assets available in your area, such as generators, medical supplies, or transportation vehicles
- Develop formal or informal agreements for sharing resources and support during emergencies, such as setting up a neighborhood tool library or creating a buddy system for checking in on vulnerable residents
- Consider creating a community resource database or online platform, where people can list and request resources, services, and assistance as needed
- Don't forget to also establish clear guidelines and protocols for resource sharing, such as how to prioritize requests, handle liability issues, and ensure equitable access

To facilitate resource sharing in your community, you work with your CERT team to create a neighborhood asset map, identifying the skills, equipment, and supplies available among residents and local businesses. You then develop a simple resource-sharing agreement, where neighbors can opt-in to share specific resources, such as a generator or a chainsaw, during emergencies. You also create a Google Doc where people can list and request resources, and you share the link with your neighborhood directory. Additionally, you work with a local lawyer to draft a basic liability waiver and guidelines for resource sharing, ensuring that everyone understands their rights and responsibilities. By establishing clear and proactive resource-sharing agreements, you help create a more self-sufficient and adaptable community that can mobilize and support itself during challenging times.

d. Advocating for Community Resilience Policies and Initiatives
- Learn about your local government's disaster preparedness and resilience policies, plans, and budgets, and identify areas for improvement or advocacy
- Attend public meetings, such as city council hearings or community forums, to voice your concerns, ideas, and priorities for community resilience
- Write letters, emails, or op-eds to local leaders and media outlets, advocating for specific policies, programs, or investments that can enhance your community's preparedness and resilience
- Don't forget to also support and collaborate with other community groups and advocates working on related issues, such as affordable housing, public health, or social justice, recognizing the intersectionality of resilience and equity

As you deepen your involvement in community preparedness efforts, you become more aware of the broader policy landscape and the role of local government in building resilience. You start attending city council meetings and learning about your community's emergency management plans and budgets. You identify a gap in funding for neighborhood preparedness programs and decide to advocate for change. You write an op-ed for your local newspaper, highlighting the importance of community-based resilience efforts and calling for increased investment in CERT training, neighborhood resource centers, and public education campaigns. You also join forces with a local affordable housing advocacy group, recognizing the critical link between stable housing and disaster resilience. By advocating for community resilience policies and initiatives, you help create a more supportive and equitable environment for all residents to prepare for and recover from disasters.

By prioritizing community involvement and resource sharing as part of your overall disaster planning, you are contributing to a more connected, resourceful, and resilient social ecosystem that can bounce back from even the most severe disruptions. Remember that building community resilience is an ongoing process that requires sustained engagement, collaboration, and advocacy across multiple sectors and scales.

As you refine your community involvement and resource-sharing strategies, consider how they integrate with your broader disaster planning efforts, such as your household preparedness plans, your workplace continuity strategies, and your local government's emergency management frameworks. By creating a holistic and systemic approach to resilience, you can help foster a more equitable, adaptable, and thriving community that can weather any storm.

PART V
RESPONDING TO SHTF SCENARIOS
IMMEDIATE ACTIONS DURING A DISASTER

When a disaster strikes, the first few minutes and hours are critical for ensuring your safety, assessing the situation, and setting in motion your emergency response plans. In the midst of the chaos and uncertainty of a rapidly unfolding crisis, it's essential to have a clear and well-rehearsed set of immediate actions that you can take to protect yourself, your loved ones, and your property. By prioritizing decisive and informed action in the early stages of a disaster, you can minimize risks, maximize resources, and lay the foundation for a more effective and resilient response.

Let's explore some key immediate actions to take during a disaster:

a. Ensuring Personal Safety
- If you're indoors, duck, cover, and hold on under a sturdy piece of furniture during an earthquake, or shelter in an interior room away from windows during a severe storm or tornado
- If you're outdoors, seek shelter in a safe and stable structure, or move to higher ground and away from trees, power lines, and floodwaters
- If you're in a vehicle, pull over to a safe location, away from bridges, overpasses, and other hazards, and stay inside with your seatbelt fastened until the immediate danger has passed

- Don't forget to also protect your head, face, and neck from falling debris or shattering glass, using your arms, a pillow, or a blanket if needed

Imagine you're at home when a massive earthquake strikes. As soon as you feel the shaking begin, you drop to the ground, take cover under your sturdy kitchen table, and hold on to one of the legs. You stay in this position as the shaking intensifies, using one arm to shield your head and neck from falling objects. Once the shaking subsides, you carefully emerge from under the table, checking yourself and your surroundings for any injuries or hazards before moving to a safer location. By taking immediate action to protect yourself, you minimize your risk of injury and increase your ability to respond to the unfolding crisis.

b. Assessing the Situation
- Once the immediate danger has passed, take a moment to catch your breath, calm your mind, and assess your physical and emotional state
- Check yourself and those around you for any injuries, and provide first aid or call for medical assistance if needed
- Survey your surroundings for any structural damage, fires, gas leaks, or other hazards, and evacuate the area if necessary
- Don't forget to also listen for any official announcements or alerts, such as evacuation orders or shelter-in-place instructions, and follow them carefully

After riding out the initial earthquake waves, you take a deep breath and begin to assess the situation. You check yourself and your family members for any cuts, bruises, or more severe injuries, and you grab your first aid kit to treat a minor gash on your arm. You then move through your home, room by room, looking for any signs of damage or danger. You notice some cracks in the walls and a few broken dishes on the floor, but no major structural issues or leaks. You turn on your battery-powered radio and listen for any official updates or instructions from local authorities. By quickly and thoroughly assessing the situation, you gain a clearer picture of the challenges ahead and can start to prioritize your response actions.

c. Activating Your Emergency Plan
- Retrieve your emergency kit and any other essential supplies you've prepared, such as food, water, first aid, and communication tools
- Activate your family communication plan, checking in with loved ones and your out-of-area contact to let them know your status and location
- Review your evacuation plan and routes, and decide whether to stay put or leave based on the specific circumstances and official guidance
- Don't forget to also secure your home and valuables, if time and safety permit, by turning off utilities, locking doors and windows, and gathering important documents and sentimental items

As you assess the earthquake damage and listen to the news, you determine that it's safest to evacuate to a nearby community shelter. You quickly retrieve your pre-packed emergency kit from the closet, along with some extra blankets, clothing, and personal items. You send a brief text message to your designated out-of-area contact, letting them know that you're safe and heading to the shelter. You also post a quick update on your neighborhood Facebook group, offering to share transportation and supplies with anyone in need. Before leaving, you shut off your gas and electricity, grab your wallet and important papers, and lock up your home. By swiftly activating your emergency plan, you ensure that you have the resources, support, and direction needed to navigate the challenging hours and days ahead.

d. Helping Others and Seeking Assistance
- If you're in a position to do so, check on your neighbors, especially those who are elderly, disabled, or otherwise vulnerable, and offer assistance or resources if needed
- Look for opportunities to share information, skills, and supplies with others in your community, such as joining a neighborhood watch or volunteering at a local shelter
- If you need help, don't hesitate to reach out to emergency responders, relief organizations, or trusted community members, and be specific about your needs and location
- Don't forget to also document any damage or losses you've experienced, as this information may be important for insurance claims, aid applications, or recovery planning

As you make your way to the community shelter, you stop by your elderly neighbor's house to check on her. You find her shaken but unharmed, and you offer to help her gather some essential belongings and escort her to the shelter. At the shelter, you volunteer to help distribute water and blankets to other evacuees, and you share information about additional resources and support services available in the community. You also take a moment to snap some photos of the damage to your home and belongings, and you make a list of any losses or expenses incurred. By actively helping others and seeking assistance when needed, you contribute to a more collaborative and resilient community response, while also ensuring that your own needs are met.

By taking decisive and informed action in the immediate aftermath of a disaster, you can minimize risks, maximize resources, and set the stage for a more effective and sustainable response and recovery effort. Remember that while it's important to have a plan and resources in place, it's equally crucial to remain flexible, adaptable, and attuned to the evolving situation and the needs of those around you.

As you navigate the challenges of a disaster, keep in mind that your immediate actions are just the first steps in a longer journey of response, recovery, and resilience. By staying connected, informed, and engaged with your community, and by prioritizing self-care, mutual aid, and long-term planning, you can help build a more robust and equitable foundation for weathering future storms.

Evacuation Procedures and Bug-Out Strategies

In some disaster scenarios, the safest and most prudent course of action may be to evacuate your home or community and seek shelter elsewhere. Whether due to an imminent threat, such as a hurricane or wildfire, or a more prolonged crisis, such as a pandemic or civil unrest, having a well-planned and practiced evacuation procedure and bug-out strategy can help ensure a smoother, safer, and more efficient transition to your next location.

Let's explore some key considerations and strategies for evacuation procedures and bug-out planning:

a. Deciding When to Evacuate
- Stay informed about the specific risks and hazards in your area, and monitor official news and alerts for any evacuation orders or recommendations
- Assess the severity, proximity, and trajectory of the threat, and consider the potential risks and benefits of evacuating versus sheltering in place
- Discuss your evacuation plans and triggers with your household members, and establish clear roles, responsibilities, and communication protocols
- Don't forget to also consider any special needs or circumstances, such as medical conditions, pets, or mobility issues, that may impact your evacuation decision and timeline

Imagine a massive wildfire is spreading rapidly towards your community, and local authorities have issued a voluntary evacuation order for your neighborhood. You've been monitoring the fire's progress and have already packed your emergency kit and important documents. You discuss the situation with your family and decide that given the unpredictable nature of the fire and the potential for road closures and smoke hazards, it's best to evacuate sooner rather than later. You also consider your elderly mother's respiratory issues and your dog's anxiety around loud noises, and factor these needs into your evacuation plan and destination. By carefully assessing the risks and benefits of evacuation, and involving all household members in the decision-making process, you can make a more informed and coordinated choice about when and how to leave.

b. Preparing Your Bug-Out Bag and Supplies
- Assemble a pre-packed bug-out bag for each household member, containing essential survival items such as water, food, first aid, clothing, and communication tools
- Include copies of important documents, such as identification, medical records, and insurance policies, as well as cash and valuable personal items
- Plan for multiple evacuation scenarios and durations, and pack accordingly, with supplies for at least 72 hours and up to several weeks if needed
- Don't forget to also prepare any necessary supplies and equipment for pets, such as food, water, leashes, and carriers, and ensure they are properly identified and vaccinated

As part of your evacuation planning, you create a personalized bug-out bag for each family member, tailored to their individual needs and preferences. For your bag, you pack a change of clothing, sturdy shoes, a headlamp, a multi-tool, and a compact sleeping bag. You also include a waterproof folder containing copies of your driver's license, passport, birth certificate, and health insurance card, as well as some cash and a list of important phone numbers. For your mother, you pack extra medications, a spare pair of glasses, and a lightweight walking cane. For your dog, you include a collapsible water bowl, a week's supply of food, and a favorite toy for comfort. By thoroughly preparing your bug-out bags and supplies in advance, you can minimize stress and uncertainty during an evacuation, and ensure that you have the resources needed to survive and thrive in your next location.

c. Planning Your Evacuation Routes and Destinations
- Identify multiple evacuation routes and modes of transportation, taking into account potential road closures, traffic congestion, and fuel availability
- Choose several potential evacuation destinations, such as a friend or family member's home, a public shelter, or a pre-arranged meeting point, and communicate these plans with your household and emergency contacts
- Familiarize yourself with the evacuation procedures and resources in your community, such as designated routes, public transportation options, and assistance programs for vulnerable populations
- Don't forget to also consider any potential challenges or barriers to evacuation, such as limited mobility, language barriers, or financial constraints, and plan accordingly

In developing your evacuation plan, you identify three possible routes out of your neighborhood, each leading to a different destination. Your primary route takes you to your sister's house in a nearby town, while your secondary route leads to a community shelter in the opposite direction. Your third route involves using public transportation to reach a pre-arranged meeting point in a neighboring city. You share these plans with your household members and out-of-area emergency contacts, and you make sure everyone has a physical map and written directions to each destination. You also research your community's evacuation resources, such as bus pickup locations and special needs registries, and you sign up for any relevant alerts or assistance programs. By planning multiple evacuation routes and destinations, and familiarizing yourself with community resources, you increase your flexibility and resilience in the face of unexpected challenges or changing conditions.

d. Executing Your Evacuation and Bug-Out Plan
- When the decision to evacuate is made, act quickly and calmly, following your pre-established plan and procedures
- Communicate your evacuation status and destination to your emergency contacts, and keep them updated on your progress and any changes in plans
- If time and safety permit, secure your home and property before leaving, such as turning off utilities, locking doors and windows, and placing valuables in a safe or hidden location

- Don't forget to also help others in your community who may need assistance with evacuation, such as neighbors, friends, or vulnerable populations, and coordinate with local authorities and relief organizations as needed

As soon as you receive the mandatory evacuation order for your area, you spring into action, following your well-rehearsed plan. You quickly gather your bug-out bags, important documents, and irreplaceable items, and load them into your vehicle. You send a brief text message to your emergency contacts, letting them know that you're evacuating to your sister's house and will update them when you arrive safely. Before leaving, you shut off your gas and electricity, lock your doors and windows, and place a note on your front door indicating your evacuation status and contact information for emergency responders. As you're driving out of your neighborhood, you notice an elderly couple struggling to pack their car, and you stop to offer assistance, helping them load their bags and providing them with directions to the nearest shelter. By executing your evacuation and bug-out plan with efficiency, communication, and compassion, you not only ensure your own safety and well-being but also contribute to a more collaborative and resilient community response.

Evacuation procedures and bug-out strategies are critical components of any comprehensive disaster preparedness plan. By proactively deciding when to evacuate, preparing your supplies and routes, and executing your plan with care and coordination, you can minimize the risks and challenges of displacement, and increase your chances of a successful transition to your next location.

As you refine your evacuation and bug-out plans, remember to regularly review and update them based on changing circumstances, such as new family members, health conditions, or community developments. Also, consider practicing your plans through periodic drills or simulations, to build muscle memory and identify any gaps or weaknesses in your strategies.

Ultimately, the key to effective evacuation and bug-out planning is a balance of preparation, flexibility, and collaboration. By staying informed, adaptable, and connected with your loved ones and your community, you can navigate even the most challenging displacement scenarios with resilience, resourcefulness, and hope.

Long-Term Survival in Catastrophic Events

In the event of a catastrophic disaster or a long-term crisis, such as a global pandemic, a major economic collapse, or a widespread infrastructure failure, survival may require sustained adaptation, resourcefulness, and resilience over an extended period of time. Long-term survival in these scenarios goes beyond immediate emergency response and short-term sheltering, and involves creating sustainable systems and strategies for meeting basic needs, maintaining health and safety, and building community resilience in the face of ongoing challenges and uncertainties.

Let's explore some key strategies and considerations for long-term survival in catastrophic events:

a. Establishing Sustainable Food and Water Systems
- Develop and maintain a diverse and renewable food supply, through methods such as gardening, farming, foraging, hunting, and fishing, based on your local climate, resources, and skills
- Implement food preservation and storage techniques, such as canning, dehydrating, fermenting, and root cellaring, to extend the shelf life and nutritional value of your harvest
- Create sustainable water collection, purification, and distribution systems, such as rainwater harvesting, gravity-fed filtration, and solar distillation, to ensure a safe and reliable water supply
- Don't forget to also build relationships and networks with other local food and water producers, such as farmers, hunters, and water managers, to share knowledge, resources, and mutual support

Imagine a catastrophic earthquake has destroyed the power grid and transportation infrastructure in your region, disrupting the supply chain and leaving communities isolated and self-reliant. In the months following the disaster, you work with your neighbors to establish a sustainable food and water system for your area. You organize a community garden and greenhouse, where you grow a variety of fruits, vegetables, and grains using organic and permaculture techniques. You also set up a collective hunting and fishing program, where skilled members share their catch with the wider community. To preserve the harvest, you build a communal root cellar and canning facility, and you teach workshops on fermentation and dehydration. For water, you install rainwater collection tanks and gravity-fed filtration systems on several buildings, and you designate a team of water stewards to monitor quality and distribution. By establishing sustainable and collaborative food and water systems, you create a resilient and regenerative foundation for long-term survival.

b. Developing Shelter and Infrastructure Solutions
- Assess and fortify existing shelter structures, such as homes, buildings, and community centers, to withstand ongoing environmental and security risks
- Explore alternative and renewable energy sources, such as solar, wind, hydro, and biogas, to power essential services and equipment
- Create sustainable waste management and sanitation systems, such as composting toilets, greywater recycling, and waste reduction and reuse programs

- Don't forget to also develop and maintain transportation and communication networks, such as trails, bikes, radios, and messaging systems, to facilitate mobility and connectivity

As the long-term effects of the earthquake become clear, you and your community work together to develop sustainable shelter and infrastructure solutions. You organize work parties to assess and repair damaged buildings, using salvaged materials and renewable building techniques like straw bale and earthbag construction. You also set up a community solar microgrid and battery storage system, which powers essential services like water pumping, refrigeration, and emergency lighting. For waste management, you build composting toilets and greywater gardens, and you implement a comprehensive recycling and reuse program. To maintain transportation and communication, you clear and maintain a network of walking and biking trails, and you set up a community radio station and messaging system using solar-powered devices. By developing resilient and regenerative shelter and infrastructure solutions, you create a safe and sustainable environment for long-term survival and community well-being.

c. *Fostering Community Health and Well-Being*
- Establish and maintain community health and medical systems, such as clinics, pharmacies, and wellness centers, to provide essential care and support
- Develop and share knowledge and skills related to first aid, herbal medicine, mental health, and disease prevention and management

- Create and participate in community education, recreation, and cultural programs, such as schools, libraries, sports leagues, and artistic and spiritual practices, to foster learning, connection, and resilience
- Don't forget to also cultivate and nurture social networks and support systems, such as mutual aid groups, conflict resolution processes, and community governance structures

As the months turn into years after the catastrophic earthquake, you and your community prioritize health and well-being as essential foundations for long-term survival. You work with local healthcare providers to establish a community clinic and pharmacy, which offers both conventional and alternative treatments using available resources. You also organize a network of community health workers and educators, who provide training and support in areas like first aid, nutrition, mental health, and disease prevention. To foster connection and resilience, you create a range of community education and recreation programs, including a school, a library, a permaculture garden, and a theater group. You also form mutual aid networks and support groups, where community members can share resources, skills, and emotional support, and you develop participatory governance structures to make collective decisions and resolve conflicts. By fostering holistic community health and well-being, you create a thriving and resilient social ecosystem that can adapt and regenerate in the face of ongoing challenges.

d. Building Resilience and Adaptability

- Continuously assess and adapt your survival strategies and systems based on changing conditions, new information, and lessons learned
- Cultivate a mindset of curiosity, creativity, and experimentation, and encourage innovation and problem-solving at individual and community levels
- Foster a culture of collaboration, compassion, and mutual support, and prioritize the needs and well-being of the most vulnerable and marginalized members of your community
- Don't forget to also maintain hope, purpose, and connection to the natural world, and celebrate the small victories and moments of joy and beauty amidst the challenges

As your community navigates the long-term impacts of the catastrophic earthquake, you embrace resilience and adaptability as core values and practices. You regularly assess and adjust your survival strategies based on feedback and learning, and you encourage a culture of experimentation and innovation, where community members are empowered to try new ideas and solutions. You also prioritize collaboration and mutual support, and you work to ensure that the needs and voices of all community members, especially those who are most vulnerable or marginalized, are heard and addressed. To maintain hope and purpose, you organize community celebrations and rituals, such as harvest festivals and solstice gatherings, and you create opportunities for reflection, gratitude, and connection to the natural world. By building a culture of resilience and adaptability, you create a community that can not only survive but thrive in the face of long-term challenges and uncertainties.

Long-term survival in catastrophic events requires a holistic and adaptive approach that integrates physical, social, and psychological strategies for resilience and regeneration. By establishing sustainable food, water, shelter, and infrastructure systems, fostering community health and well-being, and building a culture of resilience and adaptability, you can create a foundation for long-term survival and thriving in the face of even the most devastating and prolonged crises.

As you develop your long-term survival plans and strategies, remember to engage in ongoing learning, reflection, and adaptation, and to prioritize collaboration, compassion, and creativity in all your efforts. By working together with your community to create regenerative and resilient systems and cultures, you can not only survive but also build a more just, sustainable, and thriving world for generations to come.

Post-Disaster Recovery and Rebuilding

In the aftermath of a disaster or catastrophic event, communities face the daunting task of assessing the damage, addressing immediate needs, and beginning the long process of recovery and rebuilding. Post-disaster recovery and rebuilding involve a complex and multifaceted set of challenges and opportunities, ranging from restoring basic services and infrastructure to addressing social and economic inequities to envisioning and creating more resilient and sustainable futures. Effective post-disaster recovery and rebuilding require a collaborative, inclusive, and adaptive approach that engages all members of the community and prioritizes the needs and voices of those most impacted by the crisis.

Let's explore some key strategies and considerations for post-disaster recovery and rebuilding:

a. Conducting a Comprehensive Damage and Needs Assessment
- Assess the physical, social, economic, and environmental impacts of the disaster, using a variety of data sources and community engagement methods
- Identify immediate and long-term needs and priorities, such as restoring essential services, providing temporary shelter and housing, and addressing mental health and trauma
- Engage all members of the community, especially those most impacted and marginalized, in the assessment process, and ensure that their voices and experiences are heard and valued

- Don't forget to also assess the strengths, assets, and capacities of the community, such as social networks, local knowledge, and resources, that can be leveraged for recovery and rebuilding

Imagine a massive hurricane has devastated your coastal community, causing widespread damage to homes, businesses, and infrastructure. In the days and weeks following the disaster, you work with local officials and community organizations to conduct a comprehensive damage and needs assessment. You use a combination of surveys, interviews, and community meetings to gather data and stories from residents, especially those from low-income and minority neighborhoods that were hardest hit. You also work with technical experts to assess the physical damage to buildings, roads, and utilities, and to identify critical infrastructure that needs to be repaired or rebuilt. Through this collaborative and inclusive assessment process, you develop a shared understanding of the impacts of the disaster and the priorities for recovery and rebuilding, and you identify key strengths and assets in the community that can be mobilized for the long haul ahead.

b. Developing a Participatory and Equitable Recovery Plan
- Create a recovery plan that is based on the needs and priorities identified in the assessment process, and that is grounded in principles of equity, sustainability, and resilience
- Engage all members of the community, especially those most impacted and marginalized, in the planning process, and ensure that their voices and experiences are reflected in the plan

- Prioritize actions that address immediate needs while also building long-term resilience, such as providing temporary housing while also developing affordable and sustainable housing options
- Don't forget to also address the root causes of vulnerability and inequity, such as poverty, racism, and lack of access to resources and decision-making power

As your community moves from assessment to planning, you work with a diverse coalition of residents, community organizations, and government agencies to develop a participatory and equitable recovery plan. Through a series of community workshops and online forums, you gather input and ideas from all segments of the community, with a particular focus on engaging low-income and minority residents who have been historically excluded from decision-making processes. Based on this input, you develop a plan that prioritizes actions that address immediate needs, such as providing food, water, and medical care, while also building long-term resilience, such as developing community-owned renewable energy systems and affordable housing cooperatives. The plan also includes strategies for addressing the root causes of vulnerability and inequity, such as creating living-wage jobs, improving access to healthcare and education, and building community wealth and ownership. By developing a participatory and equitable recovery plan, you create a roadmap for a just and sustainable future for all members of the community.

c. Mobilizing Resources and Partnerships for Implementation
- Identify and mobilize a range of resources and partnerships to support the implementation of the recovery plan, including government funding, philanthropic grants, private sector investments, and community-based assets and capacities
- Develop collaborative and transparent mechanisms for allocating and managing resources, such as community-driven grant programs and participatory budgeting processes
- Build and strengthen partnerships and networks across sectors and scales, such as between local community organizations and national disaster response agencies, to leverage expertise, resources, and capacity
- Don't forget to also invest in building the capacity and leadership of local community members and organizations, especially those from marginalized and impacted communities, to lead and sustain the recovery and rebuilding process over the long term

With a recovery plan in place, you work with your community partners to mobilize the resources and partnerships needed to turn the plan into action. You advocate for and secure government funding for critical infrastructure repairs and rebuilding, such as roads, bridges, and water systems. You also work with local and national philanthropic organizations to establish a community-driven grant program that supports grassroots recovery and rebuilding projects, such as community gardens, cultural events, and youth leadership programs.

To support the implementation of these projects, you develop partnerships with a range of organizations and institutions, such as universities, businesses, and faith-based groups, that can provide technical expertise, volunteers, and other resources. At the same time, you invest in building the capacity and leadership of local community members and organizations, through training programs, mentorship, and networking opportunities, so that they can take ownership of the recovery and rebuilding process and sustain it over the long term.

d. Monitoring, Evaluating, and Adapting the Recovery Process
- Establish participatory and transparent mechanisms for monitoring and evaluating the progress and impacts of the recovery and rebuilding process, using a range of quantitative and qualitative indicators and methods
- Engage all members of the community, especially those most impacted and marginalized, in the monitoring and evaluation process, and ensure that their experiences and perspectives are valued and acted upon
- Use the findings and lessons learned from the monitoring and evaluation process to adapt and improve the recovery plan and its implementation, in a continuous cycle of learning and improvement
- Don't forget to also celebrate and share the successes and achievements of the recovery and rebuilding process, and to use them as a source of hope, inspiration, and motivation for the long journey ahead

As your community moves forward with the recovery and rebuilding process, you work with your partners to establish a participatory and transparent monitoring and evaluation system. You develop a set of indicators and methods for tracking progress and impacts, such as the number of homes rebuilt, the diversity of businesses reopened, and the level of community engagement and leadership in the recovery process. You also create forums and channels for community members to share their experiences and perspectives on the recovery process, such as through community scorecards, storytelling sessions, and social media campaigns. Based on the data and feedback collected through the monitoring and evaluation system, you work with your partners to adapt and improve the recovery plan and its implementation, in a continuous cycle of learning and improvement. At the same time, you celebrate and share the successes and achievements of the recovery process, such as the opening of a new community center or the launch of a community-owned solar energy project, to build hope, inspiration, and motivation for the long journey ahead.

Post-disaster recovery and rebuilding are complex, long-term, and often messy processes that require sustained collaboration, adaptation, and innovation. By conducting a comprehensive damage and needs assessment, developing a participatory and equitable recovery plan, mobilizing diverse resources and partnerships, and monitoring and adapting the recovery process, communities can build back better and create more resilient, sustainable, and just futures for all.

As communities embark on the journey of post-disaster recovery and rebuilding, it is important to keep in mind that the process is not just about restoring what was lost, but about envisioning and creating a better future. This requires a willingness to challenge the status quo, to address the root causes of vulnerability and inequity, and to embrace new ways of thinking and doing. It also requires a deep commitment to collaboration, inclusion, and empowerment, so that all members of the community can participate in and benefit from the recovery and rebuilding process.

Ultimately, post-disaster recovery and rebuilding are not just about bouncing back, but about bouncing forward to a more resilient, sustainable, and equitable future for all. By working together in a spirit of solidarity, creativity, and hope, communities can turn the tragedy of disaster into an opportunity for transformation and renewal.

Conclusion

Throughout this book, we've explored the essential knowledge, skills, and strategies for preparing ourselves, our families, and our communities to face and overcome the challenges of disasters and emergencies. From understanding the risks and impacts of different types of hazards, to developing practical skills in shelter, water, fire, and food, to building emergency kits and plans, to cultivating resilience and community solidarity, we've covered a wide range of topics and tools for enhancing our preparedness and resilience.

But perhaps the most important lesson of this book is not just what to do, but how to think and feel about preparedness. Preparedness is not just a checklist of tasks or a set of supplies, but a mindset and a way of being in the world. It is a mindset of proactivity, adaptability, and empowerment, a way of taking responsibility for our own safety and well-being, while also recognizing our interdependence and shared fate with others. It is a mindset of hope and possibility, a way of envisioning and creating a better future, even in the face of grave threats and uncertainties.

Embracing the prepared mindset means cultivating a set of attitudes and habits that can help us navigate the challenges of disasters and emergencies with greater clarity, confidence, and compassion. It means developing a sense of situational awareness and risk assessment, so that we can anticipate and mitigate potential hazards before they become crises.

It means fostering a spirit of curiosity and lifelong learning, so that we can continually expand our knowledge and skills in preparedness and resilience. It means nurturing a culture of collaboration and mutual aid, so that we can work together to build stronger, more inclusive, and more sustainable communities. And it means practicing gratitude, humility, and self-care, so that we can maintain our own physical, emotional, and spiritual well-being, even in the midst of chaos and trauma.

Embracing the prepared mindset is not a one-time event or a destination, but an ongoing journey and a way of life. It is a journey that requires courage, commitment, and compassion, a willingness to face our fears, challenge our assumptions, and reach out to others in a spirit of solidarity and service. It is a journey that will have its ups and downs, its successes and setbacks, its joys and sorrows. But it is a journey that is worth taking, because it is a journey towards a more resilient, sustainable, and just future for ourselves, our loved ones, and our world.

As we come to the end of this book, I invite you to reflect on your own preparedness journey and to ask yourself: What steps can I take today to embrace the prepared mindset and to build a more resilient and compassionate future for myself and others? What knowledge, skills, and resources do I need to acquire and share? What relationships and networks do I need to nurture and support? What stories and visions do I need to create and celebrate?

The answers to these questions will be different for each of us, depending on our unique contexts, capacities, and callings. But what unites us is our shared commitment to preparedness and resilience, our shared hope for a better world, and our shared belief in the power of human creativity, collaboration, and compassion to overcome even the greatest challenges and crises.

So let us go forth from this book with renewed purpose and passion, ready to embrace the prepared mindset and to build a more resilient and compassionate future together. Let us remember that preparedness is not just about surviving, but about thriving, not just about bouncing back, but about bouncing forward, not just about preparing for the worst, but about creating the best.

And let us never forget that, in the end, the most important resource we have for preparedness and resilience is each other. For it is in our connections and our care for one another that we find the strength, the hope, and the love to face whatever challenges may come, and to create a world of greater safety, sustainability, and solidarity for all.

So let us embrace the prepared mindset, together, and let us begin the journey towards a more resilient and compassionate future, today and every day.

www.ingramcontent.com/pod-product-compliance
Lightning Source LLC
Chambersburg PA
CBHW050301230526
45471CB00005B/1968